# tredition®
www.tredition.de

I0516222

Heiko Wenner

Bioenergetic Information Management

Managing your own power, energy and well-being.

www.tredition.de

© 2019 Heiko Wenner

Publishing and printing:
tredition GmbH
Halenreie 40-44
22359 Hamburg

ISBN
Paperback: 978-3-7497-1699-9
Hardcover: 978-3-7497-1700-2
e-book: 978-3-7497-1701-9

This work and all of its parts are protected by copyright. Any use without the consent of the publisher and the author is prohibited. This applies in particular to electronic or other reproduction, translation, distribution and provision to the public.

An Introduction into Bioenergetic Information Management

Managing your own power, energy and well-being

Heiko Wenner

Foreword

My reason for writing this book is simple: though the world often leaves us feeling small, helpless and ineffective I choose to share hope and to showcase the opportunities we have to improve and manage our life circumstances and our energy.

Based on a host of practical examples you will be filled in on the secrets of Bioenergetic Information Management.

The main reason most people barely get through the day and end up feeling, dejected, listless and demotivated is a lack of energy. An effective way of stimulating your body with energy is the use of Bioenergetic Information Management as developed by myself.

I am talking about the kind of energy that motivates you to really go out and do something, to create something, to make that change, to take control and grab life with both hands – the kind of good energy that get results you can be proud of.

Basically, the usage of Bioenergetic Information Management is so simple the many people just cannot believe it at first.

I hereby promise you that you do not have to be a physicist to understand this method and to use it for yourself.

In order to do justice to the concept outlined in this book, the book is divided into 3 parts.

Part I

Basics

Here I provide you with an introduction into the topics of energy, frequency, vibration and information based on a host of simple examples. I answer questions such as, what is health, vital energy, self-healing powers and what is the purpose immune system. In practical terms, I will instruct you on how to sense your energy, to recognise the reason for performance lapses and to protect yourself from energy robbers.

Part II

How the Bioenergetic Information Management came about

Here I outline the evolution of the energy information chips in a series of exciting reports. I also take a closer look at the multitude of healing successes and the scientific findings that contributed significantly to the methodology of development in the end.

Part III

Working with the akury Information Chips

Based on practical examples I will give you an overview of the diverse uses of the akury Information Chips and how to handle them in this chapter. We also take an in-depth look at how the akury Information Chips work, how they act and who they can benefit.

This book - ‚Managing your own power, energy and well-being' - is the result of over 50 years of research and my efforts to uncover the simple secrets of self-healing. If you have ever asked yourself the question - are we really connected with everything which exists and, if so, how can I use this realisation for myself? - then this is the right book for you.

I believe, it will take a new kind of thinking so that we can live lives full of health, joy, abundance and peace. This book will lead you to the realisation that we are all very much in a position to manage your energy, power and well-being on your own. Along the way you will also understand that the basis for deep healing and the greatest joy can be found in the ability to understand and use the principles of Bioenergetic Information Management.

Heiko Wenner
Höchst im Odenwald, October 2018

What would being able to manage your own power, energy and well-being truly mean?

Just imagine being to stay cool, calm and collected in difficult situations given a little practice? Or being able to deliver an optimal performance at any given time? You could appear confident and balanced no matter how tricky the situation and would always be in control. You could tackle your work with greater concentration and focus and be significantly more effective. Fear of any difficult situations and conversations would not be such an issue any more. After a tough day you could unwind and regenerate more quickly and also benefit from a deeper and more relaxing sleep. Wouldn't that be a game-changer?

Bioenergetic Information Management has turned all this into a reality.

The reader will be filled in on the secrets of Bioenergetic Information Management by this book and the simple practical examples it contains. It aims to understand and use the new technology. It is a description of how the method and its complexity and power are condensed into a single information chip which makes the owner's life easier.

Part I

Basics

How Bioenergetic Information Management works

Renowned scientists grappled with the terms ‚energy and vibration' over 100 years ago. Some of their findings are outlined in the following quotes:

‚If you want to find the secrets of the Universe, think in terms of energy, frequency and vibration.'
(Nikola Tesla)

‚Everything is energy and that's all there is to it. Match the frequency of the reality you want and you cannot help but get that reality. That is not philosophy. That is physics.'
(Albert Einstein)

Everything is energy, frequency, vibration and information

Even in solid matter such as stones and crystals the molecules are in constant vibration. Every thought and every emotion is accompanied by vibrations of varying frequencies. Every vibration corresponds to information, frequency, sound, colour, power and energy.

The resonance principal comes into play here in our understanding of the Bioenergetic Information Management.

Science describes resonance as how a system which is free to vibrate is forced into vibrational motion when stimulated by a second vibration. We know for a fact that all matter, even solid bodies, vibrates on an atomic level. In that way matter is able to resonate. We just have to find the suitable frequency for the respective matter.
Here is an example to illustrate: if I have the forks of our musical scale (from low to high C) in a room and I strike a second H fork, only one tuning fork will go into vibrational motion. That is the tuning H fork. Other tuning forks remain unaffected . Put simply, that means the first H tuning fork resonated with the second H tuning fork.

What do resonance frequencies do?

We all are subject to the laws of how energy is generated and consumed. The prerequisites for generating energy in an optimal manner are wholesome foods and optimal pulmonary and digestive functions. The many stimuli which have an impact on the body are pivotal in terms of energy consumption.

It can be said that we are healthy when the organism can respond to the stimuli and the baseline state can be attained again. This is only possible if the type and strength of the stimuli are such that the body can cope with them. However, each and every stimulus response consumes energy. Adapting to stimuli which are too intense, prolonged, alien and non-biological takes a lot of energy and causes impairments to your well-being and even illnesses. It is our objective to re-establish the energetic balance. If our energy field is in balance, our immune system is in a position to fight off negative influences. Our well-being and our performance are based on an energetic balance. If the energy is not in flow, disharmonious waves and blocks of the individual cells come about.

How can we benefit from this realisation?

Each and every cell, organ and organism has a specific and unique resonant frequency spectrum. A substance's wave can unleash an effect by causing another substance to resonate. (For example: the tuning fork). In practical terms, this means from the multitude of frequencies entering the body, the only ones to actually have an affect (to react) are the ones that trigger a resonance. This unique technology for the transfer information which is based on the findings of quantum physics means disharmonies can be balanced by the right akury Information Chip. The akury Information Chips are made up of a plastic platelet in shape of a square with rounded corners of approx. 20mm x 20mm. Carrier substances are applied to the surface which have an energetic charge according to the desired effect. According to the resonance principle, the client is provided with the right frequency to activate self-healing powers. One could say that harmonious information replaces disharmonious information. The duration of the process depends on type and extent of the dis-balance.

What does it actually mean to be healthy?

The definition of health as per WHO
‚Health, as defined by the World Health Organization (WHO), is ‚a state of complete physical, mental and social well-being and not merely the absence of disease or infirmity. The enjoyment of the highest attainable standard of health is one of the fundamental rights of every human being without distinction of race, religion, political belief, economic or social condition.'
(Source of reference: https://flexikon.doccheck.com/de/Gesundheit)

The definition as per DocCheck
‚Health is not a separate finding but comes under the antonym of the hardly attainable ideal ‚vital state' and its counterpart, death.'

In clinical terms, health is often reduced to physical dimensions and is understood in a simplified manner as the ‚absence of illness'. The term ‚complaint' as a concept of convenience is used to characterise the transition area between both states. On the other hand, social ethics views health as a ‚higher good' and is an ideal closely linked to the term happiness.

Regardless of the context, it can be said that health is a state which is felt subjectively beyond the realm of diagnostic verifiability. Health and illness are very much united by grey areas: one can be ill and still feel perfectly healthy– given the absence of symptoms. On the other hand, a patient can feel very ill although they are considered perfectly healthy in the clinical sense.'

How I define health is outlined by a statement of my own which some may find a little provocative:
‚Health is the current state as defined by each individual and can be influenced by the individual.'
(Source of reference: https://flexikon.doccheck.com/de/Gesundheit)

What do we mean by the term self-healing power ?

The term ‚self-healing power' can be defined as the the organism's ability to heal illnesses and traumas by oneself. (DocCheck)
(Source of reference: https://flexikon.doccheck.com/de/Selbstheilungskraft)

Our bodies posses the ability to heal themselves as long as the self-healing capabilities are not inhibited by our mental and/or emotional frame of mind.

Our organism can be said to be in balance when we are emotionally and mentally stable. This inner harmony is, however, very susceptible and can very quickly be thrown out of equilibrium through negative thoughts and feelings.

In the event of such an occurrence, our brains take on the role of a guardian and take immediate action so as to restore the sense of order without delay.

It is possible to compare our brains to the control centre of our in-house heating system. If the heating thermostat is set for, let's say, 21.5 degrees Celsius, the thermostat relays an impulse to the control centre if the room temperature drops under 21.5 degrees Celsius. Not until the room temperature has reached 21.5 degrees Celsius it has been set for, does the heating element actually stop. Analogous to the main computer processor CPU, our brains also work in a kind of review mode with every cell in our bodies. If something is not in order within the system, our control centre reacts without delay and introduces appropriate safeguards and repair actions.
If you fall and graze your knee, your brain registers this and takes appropriate actions without delay: it immediately sends white blood cells to the injured area to repel and destroy the invasive, dangerous germs. The blood vessels in the area of the injury so that we do not bleed to death and the blood clots. Who wouldn't like to have that kind of facility at their disposal?

All of these self-regulation processes happen within a fraction of a second without any contribution whatsoever from you. Truth be told, you do not even notice the process.

Our bodies are even capable of significantly more. For example, broken bones fuse together in our bodies and may even be stronger in the damaged areas than ever before reducing probability of a new break the same area.

This intelligent interaction of the protection and self-regulation mechanisms makes it abundantly clear that no body in the proper meaning of the word wants to be ill. There is an inner doctor within each and every one of us. We just have to stop blocking him or her with our thoughts.

Fundamentally, all a therapist can do is create the best possible conditions so that the self- healing powers in the body can get going. Therefore, it can be said that it is not the therapist but our very own bodies that do the healing. This also means that each and every one of us bares the responsibility for their own health.
This quote by E. Coué hits the nail on the head: ‚every illness is curable but not every ill person.'

I think the most important prerequisite for our development of self-healing powers is the conviction (belief) to get and stay healthy. In the course of my work as a healer I have accompanied many people who were considered hopeless cases- Though they had reached the end of their options with conventional therapy, they still believed in their recovery despite the severity of the illness. To the astonishment of conventional medical practitioners they recovered from their illnesses. In these cases I acted as an energy guide, impulse generator and motivator but it was not me who healed the people but the people themselves. In many cases I gave the people something to support them. Something they could hold onto which was meant to strengthen their belief in a swift recovery. Back then I developed the first information chips to strengthen the immune system and to activate the vital energy, some people wore the information chips for the entire period of illness. Whilst some people decided to put the chips aside in a special place for safe-keeping after their illness, others still wear them to this very day. Many people have taken the time after their recovery to let me know that they would not have survived without support the use of the akury Information Chips.

Strengthening the immune system is an intelligent way to stay healthy

The immune system is like a guardian that protects the body from detrimental environmental effects and is essential for the survival of the organism.

The main tasks system are as follows: to break down and remove pathogens, to recognise and neutralise contaminants and to destroy pathologically altered endogenous cells. The organism gets ill if the immune defence is weakened.

What weakens the defence ?

Along with sleeping disorders, environmental pollution and bad living habits negative stress is one of the main causes of the depletion of the immune system. Studies have shown that people who are constantly stressed get ill more quickly and are more slow cover to recover. There are all kinds of stresses such as too much work, too high expectations and time pressure. Interpersonal conflicts, separations or losses and the pressure of 24/7 availability brought about by modern technology also play a significant part.

Here it is also possible manage your energy, your power and your well-being yourself. In this case we consciously use control technology to deliver the energy to where it is needed, be it as a prophylaxis or in the event of illness. The akury Information Chips support this process by strengthening the self-healing powers and therefore also the immune system without requiring any great deal of effort from us.

My first major healing

One day in 2008 I made my way to a family in Langen to carry out a biology-oriented examination of their home. During the examination I noticed the then 16 year old daughter of the family was not well. I was able to notice this based on my experience as a healer. Her mother shared her tragic story with me: her daughter Isabelle had fallen ill with an aggressive sarcoma 2 years ago and had already had an operation. Afterwards she was given chemotherapy as well as radiation therapy. The side effects of which caused her entire motor system to fail. Her parents had to provide care for her over months and she spent her days in a wheelchair. At first the doctors gave Isabelle a chance of survival of less than 5%. Later when she believed that she had finally overcome her severe illness, there was a recurrence. Again there had been another operation and she was to return to the clinic for chemotherapy in a few days. I asked her parents to find out if it would be okay for their daughter if I were to do some energy work with her.

When scanning her body with my hands I could see a device with numerous tubes in my mind's eye. This device was to be used to administer chemo. I had never seen such a device before yet I had the impulse to give Isabelle a couple of the ‚akury eProtect' Chips I had developed. I told her to attach the chips to the bags, bottles and the device itself. My energy treatment made a noticeable difference to her and she asked me if I would accompany her throughout the rest of the process. Her parents were convinced of my healing work. With the consent of her doctors, she attached the chips to the spots I had described to her. Even though the dose of chemo was a great deal higher than the one she had been administered two years previously she has no motor problems this time. She was able to do most everything. She was even able to work on her laptop during the chemo and use the bathroom without assistance. To the astonishment of her doctors, she tolerated the therapy unexpectedly well. When I first visited her in the clinic, I saw the very device I had seen a few days before in my mind's eye.

Using my testing procedure I was able to measure the energy of the chemo and determined that vibration was minimal. Having attached ‚eProtect' Chip, the vibration of the substance was increased over tenfold. Though the contents remained the very same, the vibration of the chemo had been changed for the better by the chip making it significantly more tolerable for the patient. Along with the harmonizing affect on electromagnetic fields it had already exhibited, this was a further very positive characteristic of the ‚eProtect' Chip.
A number of my acquaintances, who are also doctors, were firm in their belief that she would not survive the illness due to her previous history. They saw no reason to get our hopes up or to offer us false encouragement. This, however, fortified me in my resolve to help her because back then I believed as I do now that: ‚even the slightest chance of survival, is still a chance worth taking because quitters never win and winners never quit.'

In the periods without chemo I visited Isabelle at home daily, to cheer her up and to motivate her to pursue the path she was already on. After the conversation, I began the healing work which took 90 minutes. We were a team and she also got homework from me to do on her own. It was my objective to strengthen her mental capabilities so that she would be in a position to deliver the energy wherever it was needed in her body and by doing so to get the self-healing powers going. Using methods I developed myself, I regularly checked to see if she had completed the visualisation, breathing and chakra exercises I had given her as well and to determine how strong her will was to get better.
Shortly before the last round of chemotherapy she hit a real low point. She questioned the very purpose of the therapy and was on the brink of giving up. During a very long pep talk I was able to bolster her morale and to motivate her to continue. I compared her situation to my experience as a seasoned marathon runner and triathlete. It can be said that a marathon does not really begin until kilometre 39. At that point at the very latest the muscles enter a state of acidosis and the pain hits turbo. No matter whether you are a pro or an amateur, between kilometre 35 and 48 it hits you like

a tonne of bricks and you start to question the whole venture. The only thing which anyone can do at that point is to rise above this weak phase using the power of your mind so that you can reach the finishing line which is so near and oh so far. I explained to Isabelle that she had already reached the kilometre 39 and had one more obstacle to overcome in order to emerge victorious and to defeat the illness. I taught her to change her attitude to chemo and to perceive it as a positive force. She visualised every drop of chemotherapy brew as wonderful potion flowing through her body and making it possible for her to return radiant health for good. The doctors in the clinic dismissed my work with a smile though they could not believe that she was doing so well in spite of the severity of the therapy.

After approximately 6 months Isabelle was ready to resume her schoolwork at home without any restrictions. Despite a break of more than a year she returned to school like she had never been away at all.

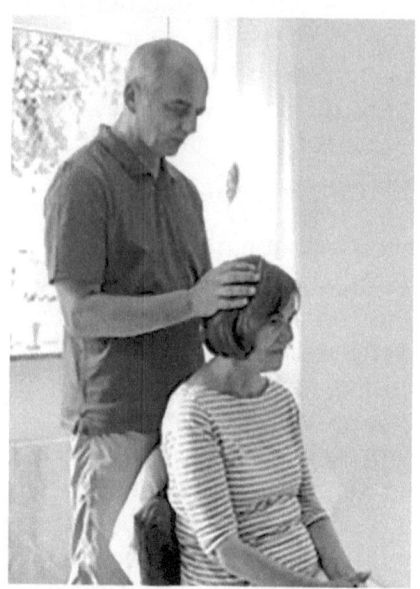

Treatment of a seminar participant

We continued to work together on a regular basis. On her next visit to the clinic and the examining professor asked her repeatedly if she, in fact, was the same person none of the medical practitioners believed had any real chance of recovery. She was pronounced cured and discharged after a severe illness to the astonishment of everyone in the examination room.

Isabelle is now a young woman of 26. After graduating from secondary school as one of the best in her class she went on to study film at the University of Mainz and then took up psychology at the

University of Würzburg. Recently she concluded bachelor studies and is now preparing for her master.

How do we perceive energy in our surroundings?

What are we actually referring to when we speak about the influence of energy on a person? How can we perceive and sense energy in our everyday lives?

Energy is everywhere and connects people, plants, stones etc. regardless of space and time. In our surroundings we sense if someone is in a good or a bad mood. Their facial expressions and gestures (shaking or nodding the head, a grimace or a smile) let us know what frame of mind they are in and we can assume how they will react in particular situations.

As a rule we sense whether the atmosphere at work is good or bad as soon as we enter the workplace. We also sense how that atmosphere affects the entire surroundings.

We perceive that the media and politics are constantly manipulating us and that spreading fear and insecurity is big business.
We sense that the ever increasing pollution means additional stress for our organism.

We sense that certain objects such as pictures or heirlooms in our living areas are of benefit or detriment to us.

We recognise that bad news and family issues or work problems can cause use to go into free-fall whereas good news such as a salary increase, a wedding, or birth can cause our energy level to rise.

How do we recognise an energy drop in time?

A more rapid onset of fatigue is normally a good indicator. We feel week, demotivated and listless. We notice that our attention span gets shorter more rapidly. If this goes on for a longer period of time, the first health issues such as headaches, migraines, cardiological and circulatory system issues, burnout symptoms and even severe depression can arise.
This may be caused by a whole host of factor which sometimes occur at the same time strengthening the impact. First of all there is the issue of the working atmosphere (the boss, the colleagues). A bad atmosphere and maybe even the fear of being fired can contribute to a high level of stress. Fear is the greatest stressor and a high stress level leads to health issues in the long run.
Our eating habits often contribute to our lack of energy. Whoever decides to make do with fast food at lunch time can expect to feel like they are on their last legs an hour later at the latest.
Many people are sensitive to the weather and react with headaches and listlessness immediately.
The atmosphere within the family can cause an energy drop just like the working atmosphere.
Bad habits such as increased alcohol or tobacco consumption. Or indeed recreational drugs can have a huge effect in terms of an energy drop.
Disruptions such as electrosmog or environmental pollutants such as an unreasonable noise level or exhaust fumes can also cause stress.

Not least because of negative thoughts and conversation, energy drops and performance lapses occur often.

Effective resources:
the energy barometer and the ‚List to live by'.

With the aid of the barometer each and everyone of us can determine their energy level.

For example:

before leaving my home in the morning to go to work, my energy level is at 8. This can be regarded as a good level. Having arrived at work, I determine that my energy level has gone from 8 to 4. There are two question to be asked here:

1. What were the triggers?
2. What can I do to increase my energy level again.

One trigger may for example be the drive to work. Perhaps it was stressful and I barely managed to dodge an accident.

Energy barometer to determine your personal energy level.

What can I do now?
First of all, I recommend compiling a personal list called ‚What is good for me? What is bad for me? In the left column you should record people, objects and events which are good for you. In the right column record what is bad for you. If I do not feel comfortable in a situation, it is expedient to avoid what is recorded in the right column and to use something in the left column as a resource. I

describe this list as the ‚List to live by' and it should be your constant companion. This also means that you should update the list at regular intervals. People who for example are not good for me at the current time may make it to the left column at some point in the future or vice versa.

What is good for me?	**What is not good for me?**
What do I enjoy? For example: reading, painting, ice cream, sport etc.	What do I not enjoy at the moment?
Which relationships energise me?	Which relationships cost me energy?
Which people support me?	Which people are not good for me at the moment?
Who do I feel accepted by?	Who do I feel rejected by?
What makes me happy?	What makes me sad?
Where do I feel good?	Where do I not feel good?
What am I drawn to?	Where do I not feel drawn to?
Which colours (clothes) do I like?	Which colours do I not like?
What do I like to eat?	Which food do I not like?

How can I protect myself from energy robbers?

Here it really is a matter of being mindful and sensing what are the triggers for your low level or my sudden energy drop and counteracting them accordingly.
As soon as you perceive performance lapses it is good to have something positive to contemplate, for example: a lovely story a picture from your last holiday. It reassures you and boosts your power to go about doing something which is good for you (the ‚List to live by'). Taking a deep breath can work a wonder. In some situations it may even be advisable to leave the room for a moment. You can, of course, also work with the energy barometer or do your own personal relaxation. Without a doubt there are many options.
For people who want to protect themselves from energy robbers I have developed a very special information chip (‚Protection and Neutrality'). Many therapists value this chip greatly in the meantime as it protects from external energies on the one hand and provides the necessary neutrality to act on the other hand.

The will to live and vital energy

It is my firm belief a person's will to live determines to a very large degree if person survives a serious illness or not.

If a person's will to live is not particularly strong or if they have quite simply lost the will to live, their chances are worse than those of a person who is ready and willing to keep going. If the will to live is lost and the patient is resigned to that fate, his chances of survival are very slim even with the assistance of cutting edge technology.

**The story one woman's death
as a result of her own mind power**

This story occurred in Switzerland when I was called on to visit an older lady in hospital. She had undergone a gallbladder operation and was doing very poorly. I visited her in intensive care with her daughter. The woman was very religious and had requested the Last Rites from the local priest because of the seriousness of her condition. She was a mother of 9 and they had all been informed of her condition. All but one of her children had already paid their last respects to their mother at this point. One son could not be located. She was very obese and even prior to the operation very restricted in her movements. When I visited her she was hooked up to a breathing apparatus and the doctors had very little hope for her recovery.

Having asked for her consent and been told that she wanted to get better, I determined that her energy level was very low and began to open her chakras and energy gateways.

Her energy level rose indicating that she had accepted the energy. In the course of the treatment a decisive impulse came to me - her husband was to apologise to her. As it turned out, her husband was, in fact, key to strengthening her will to get better and getting the healing processes going. He and he alone was in a position to do something and contribute to her recovery. As I later found out, they had had a serious argument about 14 days prior to the operation. Based on which she had decided not to speak to him until he offered an apology to her.

I spoke to her daughter and she asked her father to reach out to his wife with a heartfelt apology. This is exactly what he did and his kind words and heartfelt sentiments were recounted to me on the next day.
After two days she was feeling noticeably better much to the astonishment of the doctors and nursing staff. Having spent a week in intensive care, she was now well enough to be transferred to a normal station.

On my next visit I noticed that she had significantly more power and was able to maintain concentration with me and the other visitors a longer period of time. I sensed, however, there was something else on her mind. The idea that getting better also meant a lot of work and dedication on her part was difficult for her. She first noticed this when she received a visit from the physiotherapist after her operation. As physiotherapists are wont to do, he asked her to do some exercises. That was far too much for her she sent him away without any fanfare. Her daughter let her know in no uncertain terms that without those exercises she could not contribute to the healing process her body required and would ultimately become an invalid.

On the following day she felt considerably worse. I noticed that had lost the will to live and could no longer accept the healing energy. One day later her hospital room was converted into death chamber where she died a few hours later. In her way she had made the decision to not be an invalid. She did not want to be a burden to anyone else. From that moment when she made the decision to not go on living, her health deteriorated rapidly. For me, personally, it was a major lesson which confirmed my belief that our thoughts have the power to keep us alive but also to put an end to our lives.

The story of one man's quest for health and the will that got him there

In February 2010 a phone call from a client whose brother was laid up in hospital after a serious sporting accident. The doctor's diagnosis was paralysis from the six cervical vertebrae and they ruled out any chance of ever walking again.

My client asked me to visit him and to give him an energy treatment. She assured me that her brother was open to any and all recommendations and treatments which could improve his condition. I visited and he confirmed that he felt positive towards the idea of working with me and it was his greatest wish to get better again. I asked him how he was doing in general and he spent the next ten minutes telling me all the things he could no longer do. ‚I can only move my left hand to my left ear, I can only lift my left leg a little, the right one not at all, my right hand is numb and just this morning the examining doctor told me it will most likely remain numb...'

I could sense his resignation although he was fully convinced that he was feeling positive. The truth of the matter was that he was terrified of facing the future in his current state and had huge doubt about ever being able to manage everyday life. Needless to say, I understood his fears and anxiety but that did not stop from interrupting his remarks and asking him to describe what the situation was like when was first rushed to the clinic. He frowned and after a short period of reflection uttered a sentence which became a game-changer for the rest of development: ‚I was unable to move either my head or my arms and legs, I wasn't even able to breathe on my own.'

I then asked him to recount all the positive changes which had occurred since his arrival in the clinic. This was quite a challenge for him and a very short period of time he reverted to talking about all negative aspects of his medical condition: ‚I can only move my left hand my head, my right hand is numb and I cannot feel a thing,

I cannot even move my right hand and the doctor confirmed today that it will stay that way...'

Intuitively, I took his right hand and placed it in mind. I asked him to close his eyes and to imagine the fingers on his right hand moving. After a few seconds I felt a slight trembling and noticed his middle finger was moving. His sister who was in the room with us could not even begin to fathom it. Clear as day to see, he moved he middle finger without my help. The most important thing was that he himself felt it too.

I made it clear to him that he had to let go of his basic beliefs ‚I will never be able to walk again, I will never be able to move my right hand again, I will spend the rest of my life in a wheelchair....' Instead he was to use every single second in the clinic to picture himself moving his hands, and his arms and legs. He should picture himself leaving the clinic on his own two legs without any assistance. He should feel his sense of well-being when the moment arrives when he gets to tell the doctors, ‚I did it! You did not that I could but I did. I am walking out of here.'

Last of all I shared with him the story of the bumblebee. ‚Scientists have determined that a bumblebee cannot fly due to its aerodynamic and physical characteristics. The bumblebee knows nothing of this so never lets it stop him flying.' I shared this anecdote with him as a kind of mental crutch and gave him an information chip to help him visualise and concentrate motion sequences. With that I took my leave.

After seven weeks I received a telephone call from his sister who asked me on his behalf to come visit him again as he would to know how best to proceed with his therapy. When I asked how he was doing I received the answer, ‚my brother can walk again'.

Based on this example it is apparent that someone's will power, positive thoughts and mental strength determine what they can accomplish. In this story my role was merely that of impulse giver and motivator. It was up to the individual to make something of it.

For such situations I coined the expression: ‚getting healthy and staying healthy mean actively working on it all the time.'

Part II

The development of Bioenergetic Information Management

How it all began

One mild summer evening I was sitting outside on the balcony of my home in the Odenwald looking up at the firmament and thinking about all the things that had happened in the course of my life. I saw a time-lapse of the many stations my life before my mind's eye. It was like scales falling from my eyes when I realised nothing in my life up to that point had been in vain. Like a jigsaw each and every piece, which can be described as life experience, belonged with another piece and resulted in the whole, a clear image of the man I am today.

For more than 50 years I, Heiko Wenner, have been grappling with topic of healing energy and have proven its existence over the course of a multitude of treatments subject to scientific observation. As a health consultant, my objective has always to train people to manage their energy, power and well-being themselves using methods I have developed so they can remain healthy or become healthy again. I have developed something to support people in this quest which I call Bioenergetic Information Management. In the meantime this new method has gained recognition worldwide in the alternative healing sector.

I am aware of the fact that I am something of a pioneer in the area of ‚Bioenergetic Information transfer'. I also know that the akury Information Chips are very often viewed with scepticism at first. I am often asked the same questions: ‚do the chips work?', ‚how are they used?', ‚how did you come up with ingenious idea?'. That is my incentive for writing this abbreviated biography. I am sure that age of self-care medicine has begun because more and more people deal with themselves and their bodies in a more conscious manner.

They are looking for ways yet unknown to stay healthy or to regain their health. The akury Information Chips can support them in this matter and are a real alternative due to their user-friendliness.

Developing my sense of intuition

I recognise that the foundation for my distinct sense of intuition was already in place when I was a small child. I spent my childhood mainly on my grandparents' farm. There I learnt to handle animals and to develop and hone my fine sense. Even then I had a way of using my hands to feel which places were good or bad for me and to recognise impending danger. I shared lucid dreams with my parents. In these dreams I saw people I knew die. As a four-year-old whipper-snapper I was, of course, not taken seriously and ordered never to speak of such things. To the astonishment of my parents, all of things I had spoken of came to pass some time later. But I learnt my lesson and decided never speak of such things with them again so I kept dreams to myself from that point on.

Even as a child and an adolescent I had to be reliable and punctual because man power was much needed on my grandparents' farm. For them one principle always applied: ‚whoever is not up and ready to go by 5 AM at the latest is a slacker!' People who did not work from sunrise to sunset were viewed as ne'er-do-wells and time-wasters. A day's work had to be visible for my grandfather at the end of each and every day. Otherwise it was quite simply a wasted day. Students, civil servants and people who carried out any sort of intellectual work had absolutely no value. In accordance with this mindset, I had to spend my holidays working on the farm. In the morning I dragged myself out of bed and crawled back under the covers in the evening dead to the world. In hindsight I have come to fully appreciate my grandparents' hard work.

To this very day I admire their discipline, their perseverance and the care shown to the animals. The animals were fed each and every morning before my grandparents would even consider having a morsel of breakfast themselves. I reckon, they did not even know the word ‚holiday'. The responsibility of care for the animals and the cultivation of crops never left their thoughts. Every Sunday my grandfather would walk across the tilled fields and pray to God for a good harvest. Far away from anything like an actual church, the fields where he walked were his church. He would talk to the plants which were meant to bestow a harvest on him. Occasionally I was allowed to take part in this ritual. Being part of that never failed to impress me. I recognised what gratitude and humility mean. I clearly saw that harvest depended on the forces of nature because I had already seen with my own eyes how a hailstorm can destroy the lion's share of harvest within mere minutes.

I understood what it meant to be responsible for animals and people. I understood how very important it is to do your part reliably and punctually everyday. It dawned on me how very difficult it must have been for my grandparents to do that kind of work day in and day out as they grew older. All of a sudden I felt a sense of duty and a kind of gratitude for helping carry their heavy load. Around this time I discovered my ability to help animals and deal with human issues using energy. One day I was mucking out the cowshed when a calf came up to me and started to lick my hand. She then turned around until my hand was at certain spot where she was apparently injured. The calf remained in that spot for about three minutes before turning around to lick my hands again in gratitude and going on her way. Animals really can sense what people are are likely to be well-disposed to them.

My father also came to discover my abilities and having recognised the good it did him, he often asked me to place my hands on his shoulders after a long day's work. Overtime my curiosity grew and I understood that I had a special abilities. I began to experiment. I purchased specialist literature and began to fathom that I could develop and pass on healing energy.

It took many years for me to realise how valuable my experience working on my grandparents' farm was as it laid the foundation for my later life as a biological consultant, energy worker and the inventor of the Bioenergetic Information Management. My experience there made a huge contribution to the kind of man I am today.

How my intuition saved me from death

As a former air traffic controller in Egelsbach I could tell any number of tales but only one of them shaped my life in a very special way. It all happened in the winter of 1991. A vintage plane, a Douglas DC 3A, was standing on the apron near the fuelling station. The plane had been chartered by a film crew for a satire. The performers were all from Ruesselsheim and the South Hessen so I knew a number of them.

I went about my work in the air traffic control as usual on that foggy Sunday. A Lufthansa pilot came to the control tower and informed me that he wanted to transfer the DC 3 to Frankfurt. The weather conditions did not allow such action at that point in time. Typically, this would have triggered a discussion about the 'real' weather conditions with the pilots. Much to my astonishment no such discussion occurred. He simply asked me to let keep him posted about the weather conditions while he waited in the airport restaurant.

The film team arrived about half an hour later. The director asked me if he could film the take-off from the airstrip. For safety reasons this was normally not allowed. In this case I was willing to make an exception and assume the responsibility for the risk. As soon as the weather conditions permitted it, I drove the film crew to the runway. They were able to get the footage they required and were nothing short of over the moon. As a token of their gratitude they invited me to join them on their light over the Rhine Valley to get the final footage on the following weekend. I was pleased to accept their kind invitation.

Saturday came and I had such a sense of unease. I could not really put my finger on it but I called the director ad cancelled my participation. On Sunday morning I was on my way to Darmstadt-Eberstadt for brunch listening to the radio. A news bulletin interrupted the song and a news reader announced that a vintage plane had crashed in the vicinity of Heidelber-Handschuhsheim. They did not expect to find any survivors. A feeling of nausea overcame me. The moment I arrived at the cafe, I grabbed the phone and called the air traffic control in Egelsbach.

That day was the saddest day of my life so far. It shocked me to my core and that shock remains with me whenever I remember that tragic story. Of the 32 people on board, only 4 survived. In part severely injured. 28 people died on that morning on 22.12.1991. For some reason I had listened to my intuition and acted accordingly. Most probably it saved my life.

From then on I have always paid great attention to that inner voice and am willing to cancel plans without any ‚obvious' reason. In hindsight it has always turned out to be the right thing to do.

The extremes of a war zone

The nineties turned out to the most exciting years of my life. In April of 1992 my deepest desire to help people in need became more and more apparent. A war had broken out in the middle of Europe which would enter the annals of history as the Bosnian War. My partner at the time and I had the feeling we had to do something and in the summer of 1992 we co-founded ‚Refugee Aid Langen'. Our stated objective at that time was to support war refugees in Germany. It was possible to provide accommodation to a number of refugee families in the homes of dedicated citizens in Langen and the surroundings. Ute and I scaled-down our space requirements and provided two Bosnian families with the floor above us while sharing the kitchen.

I worked doggedly and ceaselessly for the people in Bosnia in my free time. In conjunction with the ‚Refugee Aid Langen' and the Malteser Hilfsdienst I managed the largest aid convoy consisting of 31 lorries and a load of over 450 tonnes to the refugee camps in Bosnia and Croatia. I was exceedingly proud and continued to pursue my mission of helping people.

News of the ‚Refugee Aid Langen' had spread to the German Foreign office the meantime. We were viewed as reliable partner and were granted financial support for our aid transports and projects. We worked with other aid organisations located in Germany and offered mutual support. That is how it came about that I took an unpaid leave of absence from my employer in March of 1994 to work as a project coordinator in cooperation with the aid organisation ‚Bridge to Bosnia'. The organisation ‚Bridge to Bosnia' was based in central Bosnia and arranged convoys at regular integrals transporting necessary humanitarian supplies from Metkovic to Zenica. 1994 was the main phase of the Bosnian War and the situation for the general population deteriorated from day to day as the transports were blocked or aid convoys were fired at meaning fewer and fewer of the supplies actually reached the war zones.

As the project coordinator I was in charge of team of 15 people on average and never let my fear show in precarious or unpredictable situations. Along with the daily happenings of war, I was subject to enormous psychological demands in the form of threats and attempted kidnappings. Nonetheless, it was my job to appear in control, to gain an overview quickly and to make corresponding decisions each and every day.

The events did not fail to leave their mark on me. Often I was unable to fall asleep or had nightmares. I worked extremely hard and my working day was between 16 and 18 hours as a rule. I wanted to get as much off the ground as possible in my time in Bosnia. I did not register Sundays or public holidays because war also did not. All the misery the war entailed never took a day off. My mornings I spent in the office if I was not part of a convoy along the Serbian-Croatian front lines providing the most remote villages with necessities. Sometimes the needy were waiting outside my office in rows of three to discuss their situation with me. It was clear to me that despite my best efforts I could not help everyone and sometimes I was threatened, spat at and insulted.

Regularly I felt I had reached my breaking point and seriously thought about the sense of my mission there. Over and over again I, however, managed to pull myself together and remembered my will power and mental strength. Thanks to my audacity I was able to accomplish a thing or two which had been deemed impossible by other aid organisations. That way I was in a position help people in the remotest regions of Bosnia.

After a multitude of successful aid projects I left the Bosnian war zone three months later with the knowledge and the drive to return immediately to continue to supervise some projects.

What the experience taught me

My time in Bosnia shaped me and I learnt to manage co-workers in extreme situations without letting my fear show. I came to realise what impermanence really means and how valuable it could be to help someone in need. Now I had found my true purpose and could act on this. This filled me with a special of joy. I had learnt to recognise what is truly important in life. To make do with the bare necessities in terms of my very survival. I learnt to be grateful for everything because nothing can or should be taken for granted. In Bosnia I learnt to be more mindful with valuable resources such as water, fuel, groceries and firewood. I learnt to be more patient, to take a step back and take on a new perspective, to see what is significant in each situation. I learnt to assert myself and not be intimidated. I learnt to cry and to appreciate each and every tear. As trite as it may sound, the war and all its gruesome facets, which I bore witness to, softened my heart and the way I express myself. No one can take this experience away from me. I did not see these events on TV or hear about them third parties. I experienced them for myself up close and personal. All these images and impressions have made me the man I am today.

The power of the mind

Two weeks after my return from the war zone something very peculiar happened to me. My partner convened a meeting of the refugee aid organisation. I, of course, also attended the meeting. Quite unexpectedly, no-one at all at the meeting posed a single question about my time in Bosnia, about my experience there or my impression of the ongoing situation there. It was if I had never been there. They proceeded with the usual agenda and spoke about the realisation of projects I considered important. Were my thoughts and their aura were so strong that is was unmistakably clear to the people in my close surroundings that I was not ready to talk about what I had experienced. In retrospect, I can clearly see that the power of my mind had created a protective wall around me energetically. The purpose of this wall was to protect me from queries I was not yet ready to answer. It was not until a few weeks later the people around me and the Press asked me how I had managed to survive the situation in the war zone. At that point I was, however, ready to face their questions and to make a public statement.

A matter of the heart

The war in Bosnia was over in the meantime and international aid organisations had pulled out. The people, however, were still in great need. The ‚Refugee Aid Langen' was now helping to rebuild the country. Hussein, the manager of the central kitchen came up with a very good idea. When I was back in Bosnia, he asked me if I could buy a herd of goats for the kitchen. He wanted to distribute healthy goats milk to people who were ill and produce goats cheese from the remaining milk. The idea appealed to me so on the following day I set off with Miroslav, a member of the kitchen, to buy a herd of goats. Miroslav had already spotted a possibility to buy some goats. We set off and came to a building which I had originally rented for the organisation ‚Bridge to Bosnia'. Miroslav told me that the people living in the house were originally from Banja Luka and now wanted to return to their former home. Unbeknownst to him, I already knew the family quite well as I had lived with the family of three for a longer period of time. The mathematics professor and his wife earned their living with the goats during the war.

We went to the goat shed together where it was their daughter's job to show us the goats one by one as is the usual practice at any of the livestock markets in Germany. Initially, she held back her favourite goat until her father insisted that she also let us see it. Miroslav selected the strongest animals. One of which was also the girl's favourite goat.

At this point I was treated to crash course in negotiating Bosnian style. The owner of the goats named his price. After a period of silence of about five minutes, Miroslav also named his price which was much less than the owner's price. A few more minutes went by without a word being uttered between the parties. The goat owner again named his price which was approximately midway between two prices. This occurred a number of times until the professor turned to me and said that he was in agreement. Normally, he would not have sold the goats to anyone at this price but he did it for my sake. He wanted Miroslav to know that. He also wanted him to know he knew the goats would be in good hands with me. Miroslav wasflabbergasted about this statement and wanted to know how exactly the professor had come to know me. The professor shared the story with him: the whole family was very grateful to me because I had made it possible for them to remain in the building and provided them with food supplies on a regular basis. Their appreciation had not lessened over time. Miroslav looked at me and asked me if there anyone I had not helped and did not owe me a debt of gratitude. I was quite content to leave that question unanswered.

The next day when it came to loading the goats the professor's daughter wept bitterly when we were about to load her favourite goat. I could not help but interfere and gave the instruction to wait a moment. I asked the girl if she would like to keep the goat and take it with her to Banja Luka. I did not have the heart to take her beloved goat away from her. Instead I chose to give it to her as a gift. In disbelief, Miroslav shook his head and said the whole matter was quite simply beyond him. I had bought the animal and having paid for it was now returning it to the original owner. I told Miroslav I did not need him to understand it because I myself often do not fully understand certain things I do until later. I tend to let my heart decide. This is something which gives me peace of mind. The goats loaded and our business complete, the girl came to me and planted a kiss on my forehead. I said goodbye to her parents and wished them very best of luck for their return home. This story will be remembered fondly by the family for many years to come. Who knows, maybe the girl, who is now over 25 years of age, is a woman with children of her own. And perhaps, just perhaps, she tells them the story her favourite goat.

To this very day I tend to let my heart decide in ways which often seem incomprehensible to others. For me erring on the side of generosity feels right and good and I am glad to share my time or energy work, akury Information Chips and the like.

‚The most precious things in life are not those you get for money.'

Albert Einstein, physicist and Nobel prize winner

Discovering the forces of nature as a source of energy

Even as a youth, I attempted to test my physical limitations and did a lot of sport in my free time. On occasion I would go an hour-long ride on my racing bike, then put on my running shoes for a 5km run to the lake. In the summer months when the weather was good I would go for a swim before running home again. That was in the early seventies. One day I got my brother to join me. While running, it occurred to me that could make an excellent competition. That thought stayed with me for a while. It wasn't until 1978 when I saw a broadcast of the first Iron Man on Hawaii that it all came back to me. Swimming, cycling and running were all components of the Iron Man too but the distances were longer and the order of the events was also different. It was my desire to take part in such an Iron Man competition and to be a finisher.

I was not able to turn this wish into a reality until July of 1998. My painstaking preparation for the events took place over the course of a year. The best known Iron Man triathlon in Germany took place in Roth Nuremberg and I had registered for it.

Klaus, the owner of bicycle shop in Langen, provided me with a mobile home for the event. That way I did not require any accommodation and could spend the night on location in the mobile home. Thomas, a devoted friend and companion, arrived in the late afternoon accompanied by Giovanni an excellent physiotherapist. The next day began at 6 AM the swimming distance of 3.8 km took place in the Rhine-Main-Danube Channel. The water temperature was around 19 degrees Celsius making a wet suit an absolute necessity. I found the swim easier than anticipated on that morning. I had set myself a goal of 1 hour 40 minutes but was able to complete the distance in 1 hour 27 minutes. Next up was the 180 km cycling route where the loops had to be completed twice.

The route profile was by no means easy because there was an elevation gain of 1200 m to overcome. The first leg of the tour was to

Greding, the most southern point. At kilometre 75 I was confronted with the notorious Solaer Berg with a 15% gradient. Hundreds of spectators cheered the participants on here. Still I had to avoid getting caught up in the euphoria of the masses and remained poised. Calmly, I looked at my pulse watch and made sure I did not exceed the pulse limit of 150 on these slopes. Any error could put an end to my hopes of finishing. I had set myself the goal of 6 hours and 30 minutes for the cycling part. Surprisingly, I completed this part in 6 hours and 10 minutes. Up next was a prime discipline, running. It was a bleak route along the Rhine-Main-Danube Channel.

At kilometre 20 the torture had already begun and muscle spasms were doing their worst. I was on the verge of giving up. To make matters worse, there was no shade along the way meaning I was exposed to the boiling sun all along the route. At kilometre 25 I saw the faces of Thomas and Giovanni who were there to cheer me on and I just kept going. At that point the time of 3 hours and 30 minutes I had set myself for the 42.2 km no longer mattered to me. My only goal was to reach the finishing line. That became my mantra. On that day I went up against my weaker self and looked him straight in the eye. I never stopped fighting. I never stopped believing. I never stopped running. Each and every participant knew that once you saw the church spire of Roth, the finishing line was no longer far off.

From the jetty in Roth we ran a loop through the Old Town back to the finishing line in the municipal park. Thousands of spectators lined the way and fired the participants on one last time. Giovanni ran beside me for a few metres and shouted: ‚you've made it, enjoy the final metres to the finishing line!' I was an Iron Man finisher. In 12:30:46 hours I had reached the finishing line. The time itself did not matter to me. What truly mattered to me, was the fact that I had gone up against myself and emerged victorious. I wanted to prove to myself I was in a position to accomplish anything as long as I put my mind to it. I wanted to prove something to myself. Using discipline, persistence, stamina and a great deal of mind power I had managed to complete one of the toughest competitions, the Iron Man, with one year of intensive training.

In a moment of quiet contemplation I wondered where the energy I required it to complete that feat actually came from. One day the secret was revealed to me. I went for a run in the woods in Langen. I enjoyed running there in the hot summer months not only because the trees provided me with much needed shade but because the woods instilled a certain type of energy in me which gave me a feeling of weightlessness. I could feel the energy of the woods and was glad to absorb it with each and every breath. It provided me with the fuel I needed for my competitions. The fuel to power through when others were at their limit. This realisation was my secret and something I later used as a component in my seminars. In this way participants learn to sense the energy of plants and to utilise this energy for themselves. I needed the Iron Man to gain this realisation and I'm happy to share my experience with my fellow human beings.

The necessity of letting go

My mother had undergone a large-scale tumour operation from which she had not really recovered. It was July of 2000 and her condition was deteriorating from day to day. The metastasis had spread to all her organs in the meantime. She was still perfectly lucid but according to her doctor this was also something which could not go on much longer.

I carried out my duties in Egelsbach and then drove to my mother's bedside for the night shift. I would often sit there holding my mother's hand and asking her to let go. She was still fighting the disease although she knew that the cancer had spread to her entire body and hope was all but gone. I accompanied her throughout the entire dying process and sensed that she had something she needed to tell me. In the end she was unable to do so. She knew that she had got many things wrong in my upbringing and wanted to ask for my forgiveness. She didn't have to say it to me because I was in a position to sense it. Even as a child I was able to receive unspoken words like a radio device. You could describe it as a kind of telepathy. At school I often knew what the teacher was going to say before she had even got the words out.

Nowadays I know the human organism consists of more than just matter. Basically, we are walking advertising columns comprising bioenergetic information. In my seminar the participants learn to use these bioenergetic information with the aim of managing there own energy, power and well-being.

My encounter with the Zen master and the beginning of my awakening

One day my secretary came up to me and asked if I had ever heard of Eckhart Tolle. What she told me, piqued my interest so I purchased a number of his books. My gateway to the world of spirituality was hereby blown right open. I was most eager to take in every bit of information and took up meditation. Shortly afterwards, I was given a book by Wolfgang Kopp. The title of the book was ‚Free yourselves of everything'.

The book fascinated me and I began to research the author. I found out Wolfgang Kopp was a practising meditation master and a former pupil of the late Zen master Soji Enku Roshi who passed in 1977. As his dharma successor, he headed the ‚Tao-Ch'an Centre' in Wiesbaden.

Without delay I looked online for some appointments which were suitable for beginners. The next meeting came around and I got an impression of Zen Buddhism. The Zen master Zensho a.k.a Wolfgang Kopp gave dharma lectures which I found fascinating and compelling. One day I resolved to dedicate myself to this topic more intensely and to take instruction from one of his selected teachers. Typically, one was meant to have spent at least two years with a teacher before being granted a place as a pupil of the Zen master. That was going to take way too long for me so on the next meditation weekend after my third lesson I took it on myself to introduce myself to the Zen master. I wanted to know if I could become his pupil. No sooner said than done. At that point he already had 109 pupils. Following his dharma lecture there was a short intermission. I seized the opportunity and knocked on the door of the meeting room. The Zen master asked me in and offered me a seat. He asked me what it was I wanted and I answered from the bottom of my heart in a loud voice: ‚I would like become your pupil'.

His deep blue eyes beamed and he answered: ‚this much I know, this wish has come from the bottom of your heart.' He asked me if I realised what I was getting myself into. I replied that it couldn't be all that bad as the other 109 pupils had managed to stay with him for a number of years. He smiled, took his bell in his hand, opened the door and rang a few times. When all his pupils had turned to him and the room was perfectly quiet, he introduced me as his new pupil. Everyone clapped and expressed their congratulations. Many of them asked me how I had managed to be granted a place as a pupil after such a short period of time and I answered: ‚you will have to ask the Zensho in person what made him do it.'

Two or three times a week I drove to Wiesbaden and took part in a monthly meditation weekend. As a newbie I got to know the customs, conventions and the members of the sangha. In Buddhist terminology sangha means ‚assembly', ‚amount' or ‚community'. Time went by. The master's dharma lectures repeated themselves and I grew bored. Within a few months I had already learnt a lot but gradually I had the feeling that it was not going anywhere and there was no real progress any more.

A message from a red slug

During this monotonous period I also sought further training from a shaman. On a number of occasions I went to Hermann Strohmeier in the Teutoburg Forest to learn shamanic medicine. In addition, I learnt about the sweat lodge ceremony and how to construct such a lodge. Hermann taught me a lot about native American rituals. In his presence I always felt secure and accepted. Besides the drum rituals, the sweat lodges, and the shamanic healing I also trained my sense of mindfulness and was able to perceive the energies in my surroundings more and more consciously. I honed my senses in the great outdoors and recognised thereby how very quickly primal forces work and how everything at the end of the day is connected as one creation.

Recognising the interaction the forces of nature was a very valuable experience for me. I would often sit wrapped in a blanket on a glade under a starry sky pondering the endlessness of the universe. I sensed the interaction as planetary forces and began to consciously use their energy for myself. I directed the energies to the painful spots in my body and felt such relief that I anchored this positive feeling there. To anchor something meant saving the information in that part of the body so as to be able to retrieve it at a later point. In a way it can be compared to how a Word document is saved on a CD, USB stick or hard drive.

At night-time I experienced moments of inspiration. My awareness of my body and my sense of intuition went from strength to strength. It became clearer day by day that there was more to this life than I had been able to fathom up to that point. Hermann had taught me that the answers to pending crucial life decisions were best found by posing clear and purposeful questions. These questions took the form of a prayer to the forces. A mindful perception of goings-on in nature could provide the answer.

I had a serious and far-reaching decision to make. The question whether my job as a commercial director in Egelsbach was the right job for me was on my mind day and night.

There was no joy left for me in my job. The expansion of the airport was in the final stage and I had accomplished everything I had set my mind to. All my projects were more or less complete. I saw no reason to carry out the unchanging routine of work day in day out. That question was on my mind when I took to the great outdoors one Sunday morning.

For some inexplicable reason, I did not see single person although it was around 10 AM. I passed through farmsteads where no one was to be seen. I walked through a residential area and it was like a ghost town. Was I in another dimension? I had no idea but the whole situation just seemed very strange to me. I drifted along feeling the warmth of the breeze on my skin and breathing in the fragrance the first blossoms of the spring. I had lost all sense of time and passed through the landscape in a meditative and mindful state.

At some point I came upon a stream. I sat down on a tree trunk enjoying the spring and the murmur of the water. I closed my eyes and contemplated the question as a prayer to the cosmic forces requesting an answer. Up to that point I had not received any information. I directed my full attention to the sound of the flowing stream and felt its energy. I had the intuitive feeling that the sound of the water wanted to tell me something I fell into a trance. In this trance I received the message that I should do something differently: ‚be as playful as the water, change and bring forth something new'. I asked when the change was to occur and then fell asleep in the midday sun.

When I opened my eyes, I saw a red slug slowly creeping over my right shoe. That was the message. The slug symbolised my slow pace i.e. I was meant to allow myself time for the process of change and not rush. I had received the answers to my questions.

In the following weeks I remained attentive but received no further information.

The instruction I received from Hermann and the experiences that went along with it were very valuable to me. So much so that I incorporate them in my seminars. I learnt to find the answers to life's grand questions in nature using mindfulness combined with my gifts as an observer. This has became a meaningful component of my life whenever I look for a solution to resolve any issue. I turn to nature with a clearly formulated question. Sometimes, it takes a while for the answer to come to me but it has not failed me yet.

While I was a pupil to the Zen master, I could not help but recognise that in each and every person there is a master. That means that I too am a master at the end of the day. This realisation was a significant factor in my decision to terminate my training with the Zen master after year and a half.

Light nourishment

An old friend of mine by the name of Regina invited me to a meditation weekend in the Ruhr region two months later. We met up at her place in Vloto and travelled to the event together. On the way Regina said she wanted to take a short detour to visit a friend of hers. This friend had been living on nothing but light for a number of months. Allegedly, there was not a morsel to be found in her fridge. I was quite dumbfounded and imagined Regina's friend to be a rickety, emaciated scarecrow of a woman. When she did, in fact, open the door I was in for quite a surprise. There she stood and was quite the opposite to what I had expected. She looked well nourished and astonishingly healthy. This really piqued my interest and I wanted to know everything I could about light nourishment. She was, however, not interested in telling me about it but gave me a number of books about this topic instead. The meditation weekend turned out to uninteresting for me as it reminded me of my time with the Zen master Zensho.

Back home in Langen, I picked up the book about light nourishment with great curiosity and began to read it. After the first twenty pages I put it down and came to realise the process just was not right for me.

There weeks before Easter I began to fast as I had the feeling I wanted to detoxify and cleanse my body again. About one week into my fast, it dawned on me that I had the books about the light nourishment process and I decided to give them a second try. This time the topic really grabbed me and I resolved to go through with the process. Easter seemed a good time so and I took an additional two weeks off.

At 5 PM on Good Friday I began the light nourishment process. I had prepared my body for the process by fasting for a period of 3 weeks. I lit a candle and began the light nourishment process with a meditation. All that remained was to do without food and liquids for the next seven days. I was already used to not eating but not drinking over a period of 168 hours was something completely new to me. Jasmuheen, the writer of handbook for the process describes that in the night of the 5th day the spiritual realm allows an energetic shift to take place to relieve the strain on the organs. A person can survive up to four weeks without eating according to experts. Without consuming liquids, however, the dying process normally begins as of the 4th day.

As of the third day my kidneys did, in fact, begin to hurt. Something must have happened on the fourth night which I did not notice because from that day on I had no more pain and felt noticeably better. Over the course of the first seven days I meditated several times daily. I even went outside for walks. I still felt very tired. I had not eaten anything for over four weeks. When brushing my teeth I had to take care not to swallow any water. Otherwise the process would have been interrupted for 24 hours meaning I would have to extend the process by one day. The objective of the light nourishment process is to re-programme the cells so that on conclusion of the process the physical body can obtain nourishment through light (biophotons). Over the course of the first seven days I truly did experience something of an emotional roller-coaster. I felt that my sense of perception was changing and I began to get to know my body in a completely new manner. My trust in the spiritual realm went from strength to strength. A sense of basic trust came about all of a sudden after the 5th day.

I was conscious of the fact there was an energy keeping me alive. I was now in a position to encounter this energy in a manner I could feel. With great reverence and gratitude to the spiritual realm I lit a candle at the end of the first seven days at 5 PM. and then began the second phase of the process.

I was now allowed to consume liquids. I attempted to drink some water but half of it streamed out of my mouth. I asked myself: ‚what is up here?' I had to re-learn how to swallow and made my next attempt with a teaspoon of water which I dripped onto my tongue. It took me more than half an hour to drink a glass of water. It became abundantly clear to me why people who have been tube-fed over a longer period of time find it so difficult to swallow again. As of the 8th day my energy level and strength really went through the roof. It had become clear to me how much energy even a glass of good water contained. In the following days I sensed how my body was returning to strength although I still had not eaten anything.

On the 15th day I returned to work and could not help recognise that my perception, mindfulness and ability to concentrate were significantly better. I perceived the energies surrounding me in a different way. It gave me a great deal of pleasure to recognise that I could sustain without nourishment. My primal fears about not getting sufficient nourishment lost all meaning to me. I had now done without any food for over 6 weeks and had more energy and power than ever before.

I also became more and more conscious of the benefit in terms of time. Over the past weeks I did not have to buy any groceries. I was able to get up later as I did not have breakfast. I also did not give any thought to what I should to eat. All in all, I was saving time and money.

The fact that I was doing without food was something which was of interest to some people. I explained the light nourishment process I had gone through to those people. Most of the time I was confronted with somewhat condescending smiles and the question: ‚if it is as simple as all that, why are there still people in the world who suffer from famine?'. My answer was quite simple: ‚because they do not have the consciousness for that.' It would, in fact, be possible to help many people by changing their consciousness and I can clearly picture how gigantic industrial sectors would collapse or would have to adapt to the monumental changes if a large of portion of mankind were to choose the light nourishment process. Supposedly, a total of 20,000 people are meant to have gone through the process successfully worldwide so far.

I did without food for more than 16 weeks and felt very good. As of the sixteenth week I began to eat again. The transition to solid food was not difficult at all. I had no digestive problems whatsoever. It became clear to me that I could do without food at any time because my cells were now programmed for light nourishment. The advantage of this was something I came to recognise over time when, for example, on a mountain hike my companions had to stop at a mountain hut for a snack after three hours at the latest. If by chance the mountain hut was closed or the way to the next cabin was to take two or three hours, some of them got aggressive or were really depleted. For such ‚emergencies' I had a supply of muesli bars for my companions. Personally, I did not require anything to eat for days on end on such tours.

There was, of course, a reason I began to consume solid food again. Eating and drinking are of great significance within our society and a part of how we socialise in general. Many contracts are finalised over dinner. As a commercial director I often received visitors during the planning of the airport expansion in Egelsbach. These were visitors from the Ministry and the government Praesidium so there were all kinds of politicians, planners and government officials. Most meetings took place before noon with the intention and hope of rounding off the meeting by inviting the guests to lunch the airport restaurant. Initially, I was able to excuse myself by claiming I had an upset stomach. That way I avoided eating and having to tell the guests about the light nourishment process. Most probably they would have considered me a nutcase or a bit of a freak. So I began by having a soup as a starter so as not to stand out or to have to answer questions I considered inappropriate for the workplace.

In terms of my own spiritual direction, this process was a significant step forward. It provided an important jigsaw piece. It was not until I was in the middle of the light nourishment prices that I became aware of the energy keeping me alive. I was now closely connected to this Divine energy and felt it clearly. I felt a lot more sensitive and permeable but also more fragile. After the light nourishment process I meditated every day and developed a method to help me connect with the Divine energy quickly. I refer to this procedure as ‚going online'. Over time I have refined this technique and have designed it to be so simple that anyone can pick it up from me in a seminar.

The medicine man's chain

My gaze was fixed on the final landing planes of the day when my secretary knocked on the door to wish me a pleasant evening. Off the cuff she asked me when I had last been on holiday and that question really hit home. I could not remember the last time I had been on holiday. It was mid February and I had no idea where I could go on holiday at this time of the year. In Europe it was far too cold as I really wanted a sunnier climate for my holiday destination. No sooner said than done: the very next day my secretary presented me with an excellent offer for a trip to Senegal.

Two weeks later I was sitting on a plane on my way to Dakar. My destination was the club Aldiana which was located near the Senegalese city of Mbour. After a bumpy two-and-a-half hour bus ride we arrived at the club. On my arrival I received a tour of the delightful holiday resort surrounded by ancient Baobabs. An obligatory part of the tour was a very clear security briefing. No guest was to leave the club without personal security. I was staying on my own in a rondavel near the ocean and after a restful night decided to disregard the security briefing and take a walk into the savanna with only my little backpack.

It was the end of the dry season and the landscape was brown from the sun and bone-dry. The puddles and watering holes were dried out. After an hour and a half of marching through the brutal heat I arrived in a small village. The village was made up of a handful of huts. These huts were made of clay and a colourful other materials best described as improvised. In spite of the intense heat, the natives were sitting in a circle in the middle the village. They sat there playing cards fully exposed to the blazing sun. It was midday.

An old man glanced at me briefly with eyes which were alert and sparkling. Our eyes met and I knew that from moment on he was someone who carried something mysterious within. He wore a cowl as a robe and seemed to me to embody pride and integrity.

A young man jumped up and approached me. It was taking us a few moments to determine what the common language for our conversation would be until he asked me in surprisingly excellent German what exactly I was doing there. I was taken aback but later found out that the club Aldiana was the largest employer in the region. A prerequisite for any kind of employment was the knowledge of two languages at least. Consequently, many of the natives were able to speak a number of foreign languages as well as their native tongue. I told him that I was an inquisitive man who was eager to learn. Specifically, I wanted to get to know the people and country. Then it was my turn to ask a question and I inquired about the old man wearing the cowl. My premonition turned out to be true. The young man answered me with visible pride that it was his grandfather and you was a well-known healer and advisor.

I asked him what exactly his grandfather healed and was told that the people of Senegal were very poor and without money for medication or doctors. For that reason, many ill people came to him and then he went out into the savanna to gather medicinal herbs. His work as an advisor also very impressive. Whenever a government official required a confirmation for resolutions pertaining to the future they came to him. He consulted his ancestors and gave a suitable recommendation.

He referred to his grandfather as Abdou. I now asked the grandson to ask his grandfather if I could accompany him the next time he went out into the savanna to gather herbs. The young man was quite horrified at that notion. To that very day his grandfather's very own sons had not been allowed to accompany him on his herb gathering trips. Now a perfect stranger was asking to do that very thing. Abdou looked at us as if has understood what was going on from afar.

Throwing caution to the wind, I asked his grandson to ask him anyway. He went to his grandfather and came back with an answer of sorts. His grandfather wanted to consult his ancestors and I was to return on the following day. I returned to the Club and waited full of suspense for Abdou's answer. The next day I set out for the little village and was able to find it right away.

Abdou and his son had been expecting me. My excitement was beyond measure. His grandson did me the favour of translating and told me that his grandfather had spent half the night consulting his ancestors and had been granted permission to take me with him the next time he went to gather medicinal herbs in the savanna.

Abdou, healer and adviser in Senegal

This was something very special to me. Two days later the medicine man, his grandson and I set off for the savanna in a carriage with no suspension drawn by a nag which had definitely seen better days. It struck me that throughout the ride that I always knew in advance where we would stop next.

After approximately two hours I glimpsed a greenish luminous tree in the distance which stood out in the dry savanna. I pointed in the direction of that tree. Abdou smiled and his grandson translated that we were, in fact, on our way to that tree. The closer we came, the more apparent it became that this tree was a majestic and powerful being of nature born of the savanna. The fact that Abdou's father had been interned under the tree made the whole thing even more impressive to me. We took a break and Abdou asked to be left alone for a short while. He meditated in the shade of the tree for a spell and then asked me to sit next to him on the ground. From his brown cowl he took a small leather bag. According to his grandson, the bag contained dried berries from this tree. Abdou handled them with dexterity and put them in three rows. After this ceremony his eyes were shining like sparkling diamonds and he confirmed that his ancestors were happy that I was there.

We gathered various medicinal herbs on that day and I was able to learn a lot from Abdou. We returned to the village in the late afternoon. All the while Abdul was wearing a chain around his neck which he then returned to its rightful place in his hut as soon as he returned.

He explained to me that he always wore the chain whenever he went out to the savanna, meditated with his ancestors or carried out sacred ceremonies.

He then invited me to dinner. The women of the village had a large dish of rice and goats meat. Abdou told me to return on Saturday for he wanted to take me to the market in Mbour to show me medicinal herbs unavailable in Senegal but in the neighbouring Gambia. I promised to return.

The natives working in the club find a nickname for me in the meantime. I was known as, the white man with the ‚red backpack'. To the other club guests I was known as, the one doesn't play by the rules.

I met on Saturday and we set out for the market in Mbour. As always I was wearing my trademark red backpack.

When Abdul was standing in the middle of the marketplace and asked me to take off my backpack, I gave him a questioning look. He took my bag and opened it. What happened next came as quite a surprise. He took off splendid chain, spat on it three times and threw is in the open backpack. I was exceedingly surprised and speechless. I stood there in front of him and had no idea what to do or say.

It was clear to me that this chain was of great value to him so I let him know there was no way I could accept such a precious and unique gift. He countered that the chain was the very reason I had come from Europe. I was supposed to keep the chain as a treasure and whenever I had a problem to be solved I was to wear it and the problem would solve itself.

This gesture brought tears to my eyes. After walking around the market with Abdou I said goodbye and returned to the Club. Full of reverence I locked the chain in safe of my rondavel.

The mystical chain of the medicine man

I spent the next days engaged in the activities offered by the club such as archery and drums but Abdou and the chain never left my thoughts.

Two days prior to my departure, I set off for the village yet again and everyone was assembled there as if they had been waiting for me. I was asked to take a seat in the assembly. Abdou's sons were present this time including one of his sons who worked in the club and spoke German. I began by saying that I had come to say goodbye as I was leaving on that day. I also wanted to have some information about Abdou's chain. No sooner had I said the word ‚chain' than all eyes were fixed on me. The group became very animated, Their gestures and arm movements demonstrated clearly they were quite furious. They spoke to Abdou in their language with great haste and agitation. His grandson told me that they could not understand why he had given me the chain which was normally to be given to his eldest son on his passing. More importantly the chain was meant to stay within the family and the tribe.

In his family's presence, Abdou confirmed that his ancestors had granted him permission to give me the chain. It had taken Abdou over 17 years to craft the chain made up of seventeen partial extremities bound in leather. Each part of the chain was made of something of great value from nature or a powerful animal. In total awe and full of gratitude I bowed to Abdou and his family. He was proud, sincere and honest man. In the meantime he had reached the ripe old age of 70 years. That age was remarkable for Senegalese standards where the average life expectancy is 59 years old. Abdou embraced me and said in broken German: ‚you take good care, my children not good. Only want money. His parting words were: ‚take care my brother!' With tears in his eyes he turned away returned to his hut without once turning back.

In hindsight, the whole thing seemed like a movie to me. My part in the movie had been clear from the very start. I felt the whole thing had been pre-determined somehow but I had no idea what my time with Abdou could mean for the rest of my life.

Discovering energy transfer

Three days later I was sitting at my desk at the airport and felt that something had changed within me. In my lat in Langen I had set up a meditation corner and followed Abdou's instructions always wearing the chain whenever I had to solve a problem, I wanted to help people or I needed protection.

A number of months had passed and I felt the need to take part in a seminar for healing work in the Eifel. It was autumn and I was wearing my chain under my pullover to protect myself from external energy. We were practising chakra work when the participants noticed that the energy around me was extremely high. At that point I offered them no explanation whatsoever for this.

At some point during the seminar I had a vision. This vision came with clear instructions. I was to hold the chain in my left hand and a quartz stone in my right hand. Using my body I was to transfer the energy from the chain to the stone.

The next day, on waking I followed the clear instructions and took the chain and the quartz stone into the seminar. I told the other participants about the chain and asked them to measure the chain's energy and to record it on a piece of paper. The chain had enormous energy (in excess of 240,000 Bovis units). There was little to no disparity in the measurements. Afterwards, the participants measured the stone's energy and to everyone's astonishment the stone had an energy of 240,000 Bovis units.
My system of energy transfer was born!

The physicist A. Bovis developed the biometer. The original purpose of the Bovis unit was to determine the ‚biological quality' of foodstuffs, materials and places. The table is used as a tool whereby the quality of an item and its vital force are determined using an antenna or pendulum. The neutral point is at 6,500 BE. Values below this are said to be energy depleting for the human organism. Values above this are said to be energy-boosting.

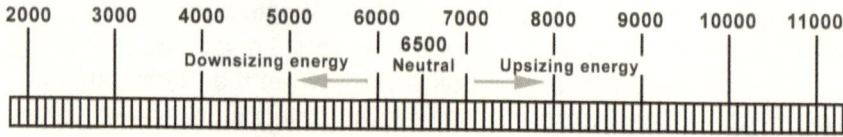

Measuring scale of biological according to Andre Bovis.

Going new ways as a building biology consultant

Day by day I became more and more conscious of the fact that it was not my life purpose to work as a commercial manager. I often sat in my office, lost in thought, looking out the window and thinking about my encounter with Abdou. I still, however, wanted to finish my project at the airport. The project took about a year until it was finally done. That stage of my life was then well and truly over and I gave notice. Now I was in a position to start something new. I wanted to pursue my true purpose at last and help people.

My friends and acquaintances could not fathom why I made such a decision and they found the change nothing short of disconcerting. It was totally and utterly beyond them why I would want to give up a stable job. When I left my job, I did so with a good feeling because on an intuitive level I knew it was for me to take a different path.

A few months later I attended my first training courses to become a building biological health consultant and founded the company ‚Building Biology Wenner'. It did not take long for the first customers to appear and my healing work was more and more in demand. My old circle of friends faded into the background but at same time more and more like-minded people entered my life.

Not only did I really enjoy the building biological examinations but I was able to help my customers quickly with simple solutions. Many of them suffered from chronic pain or insomnia. By implementing the measures I recommended, many issues were easily improved. First and foremost, the examination involves recognising influences which are detrimental to health in the surroundings, balancing such energy fields and activating available energy potentials.

This optimisation is of particular importance for people whose health is already impaired. Over the years I have developed my own examination methods which consists of four parts. Put together, these make up a jigsaw to represent the whole. This whole shows me the

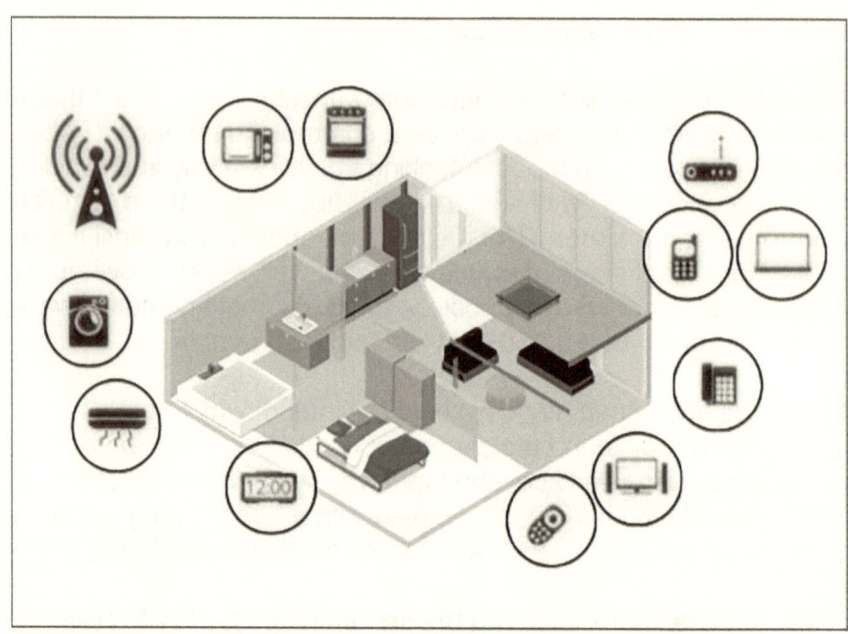

Possible source of electrosmog in residential buildings

problem areas to which I can provide an appropriate customised solution. I coined the term ‚the four components examination' for my method. It consists of the following parts:

1. Electrosmog pollution
   Unbeknownst to us, we are constantly exposed to electrical and magnetic fields as well as electromagnetic waves. With my technical devices, I check for electrosmog pollution in the bedroom and the workplace according to the building biological guidelines of the high and low frequency.

2. Geopathic pollution (Earth radiation)
   Earth radiation, water veins, faults and global grids may be the cause of health problems. A harmonised bedroom and workplace are the basis for vitality and well-being.

3. Contaminants Contaminants in construction materials and decor can trigger numerous illnesses.

4. Negative influences Objects, heirlooms and messages can damage our energy system or unconsciously.

Using the ‚four component examination' I have been successful to this very day and have used it on more than 2,000 occasions in living and work spaces as well as public buildings.

Harmonization of earth radiation by a stone circle

Over the years I have developed a system to harmonise earth radiation. Not everyone has the possibility to avoid earth radiation in their living space so I developed a method to harmonise bedrooms and workspaces without having to move the furniture around. To do so, I use an ancient technique known in Neolithic and Megalithic cultures. Back then, the priest known as druids protected their knowledge about the power of stone circles like the Holy Grail. I visited a vast number of these stone circles and got to grips the wisdom of the druids slowly but surely. I recognised the system used to the stones. I sat in the middle of the circles and sensed there magical energies. It felt like I had been transported to an earlier time. I received the images and information I required to replicate the stone circles on a smaller scale. Over the past years I have often wondered if perhaps I had been a Druid in a former life for the information I received seemed to me more like a reactivation of old knowledge rather than completely new knowledge.

I recognised that the stones were more often than not quartz and because of that were able to store the information which had been imprinted on them to this very day. Silicone is, for example, the main component in our modern storage media. The quartz boulders of the stone circles were mostly of granite. I found out that the plus and minus powers of two stones were always perfectly aligned. It is comparable to two rod magnets where plus and minus polls are put together. The shorter the distance between magnets, the stronger the power of attraction becomes and at some point an energy field comes into being. If, however, both plus polls or both minus polls are put together, they repel each other. This means that all stones must be perfectly aligned and the distance between them correct so that the harmoniously acting stones bring about a highly oscillating field. You can compare it with an orchestra in which all the instruments are in harmony and result in a symphony. I use this technology today to harmonise living and office spaces, hotels, schools, kindergartens and clinics.

It is clear to me that I received all this information from a very large field of consciousness so that there are actually no secrets. People who are in a position to connect with this field of consciousness can receive all the information required to resuscitate long lost knowledge or to push new developments. I make no secret of that fact and am delighted to share my knowledge with the participants in my training courses.

My form of building biological examinations represents a significant element in the observation of the all-encompassing, complex, humanistic big picture. Before I begin my energy work with severely ill people, the disruptive factors in their surroundings must be eliminated. Otherwise it makes no sense whatsoever because that newly found energy will go up in smoke in no time.

In order to sense the negative influences I connect with the energy field of the affected person and obtain further information about them.

The message of an old stone jug

Whenever I examine someone's living space, I also look for any negative factors which could influence that person subconsciously, provided they have no objections. I pose myself a question mentally: what is it in this person's surroundings which is not doing them good? I use my special antenna to detect negative influences.

This antenna was activated during a building biological exam when in direct proximity to a very old stone pot. The stone pot was located in the corner of the living room. I asked the owner if she had inherited it and if perhaps some ill will from certain members of the family was somehow connected to this heirloom. She told me, the pot had belonged to her grandmother but she knew of no envy, jealousy or ill will from any of her relatives based on the fact that she had inherited the pot. I asked her to give it a while as I continued to do my work. A few minutes later, she told me that every Friday when the cleaning lady was in the household, she was filled with a sense of apprehension that something could happen to the pot. After this statement, it was crystal clear to me that the fear of losing something valuable was depleting her energy.

Again I posed myself the question mentally: what is a good place in this household for the pot? My antenna drew me to the next room where there was living room cabinet. The right place for the pot was on top of this cabinet. The owner looked at me quite dumbfounded and could not hold back her tears. This cabinet was also an heirloom which had belonged to her grandmother and back when she was a child the original spot where the pot stood was the very spot I had suggested.

The search for a gold object

When I had concluded the building biological examination of the living room and was discussing further steps the owner, the owner's husband arrived home from work. As an IT specialist, he could not have been more sceptical about my work and referred to me as a quack. His reasoning for doing so was simple: nothing I did could be scientifically proven.

I remained calm, took a deep breath, took my antenna in my hand and asked him if there was a golden object anywhere in the flat. He confirmed this and I attuned myself mentally to a golden object. My search did not require any great distance because the antenna was activated by the sideboard right in front of him. When I asked him to open it, he swore solemnly there was no way in the world we would find a golden object in that sideboard. I asked him to open the doors anyway. My antenna was activated by a compartment in the lower part of the sideboard where the silverware inherited from the grandmother was located. I asked him to open the case of silverware. To his astonishment, a small golden spoon was in the middle of compartment. Surprised and embarrassed, he apologised for his behaviour more than once and thanked me for convincing him. In jest I asked him if I could continue my search for gold objects within the flat keeping all the objects which in his opinion ‚could not be gold'. With a smile he rejected my kind offer.

This story outlines how it is possible to consciously connect to certain energy fields to receive corresponding information.

Obtaining information from a field of consciousness

Subsequent to a lecture in Stuttgart, an old lady came up to me and asked me to take a look at her as she had been feeling unwell. I asked her to take a seat and determined that she was ‚down' in terms of her energy. I resonated with the vibration (informational or mental field) and received information ‚asparagus'.

I asked her if she had eaten asparagus in the last few hours. She confirmed that she had eaten asparagus on the previous evening and it had triggered a reaction. The reaction comprised a reddening of the face, instability her circulatory system and a lack of energy. In the end, it dawned on her that she has had the very same issues last year at the same time. Back then the reaction had been so serious that she had been taken to hospital where no doctor was able determine what the cause was.

So now to the key idea of this story: how did I obtain the information from the mental field?

The participants in my seminars get to know a multitude of exercises so as to resonate with their surroundings and/or with another person relatively quickly to obtain the necessary information.

By the way, I have coined a simple term for resonating with something. I refer to it as ‚going online'. In the meantime most everyone understands the concept of going online as it pertains to technology and how we obtain information from the internet. To obtain information from the internet I have to go online and then obtain the desired information using a browser.

‚Going online' with our surroundings and/or a certain person we would like to resonate with works in a similar way. One prerequisite is that a connection exists. This connection can only come about when both, transmitter and receiver, are willing to work on the same frequency. That means it is no use trying to listen to a certain station on a radio device if the frequency is set to a different station, or that station is not on or is somehow out of order.

'Going online' also means connecting with the energy of the source of all things or the all-encompassing Divine energy and blending with this energy, true oneness. The Divine energy and/or the Divine spark is in everything which exists and has a vibration and therefore a frequency. That is the only secret and it makes apparent that each and everyone can connect with everything which exists if they are not given to doubt about their abilities to do so.

A dispute and its consequences

Some time ago, the mother of an eight year old girl called me and told me that over last 3 weeks her daughter had been waking up at night and coming to her bed full of anxiety, trembling and weeping. She had no idea why this was happening and asked for my advice. I connected with the parents mental field and visualised the following which I then described to the mother: ‚just imagine sitting at home with your husband and daughter at supper and suddenly there is discussion which ends in an argument. As this has happened number of occasions, you decide to face facts and shout loudly: ‚if this happens again, I am going to pack my bags and leave'.

There was silence on the other end of the line. Not until I asked if she was still there did I receive an answer: ‚how do you know all that? That is exactly what happened.'

I asked the mother if she could begin to imagine how her daughter felt in this situation and described to her what daughter may have thought and felt in that moment: ‚my parents are fighting and that is not good. My mother wants to leave me. My mother does not love me any more. Now I have to take care that she does not leave me and abandon me.' The mother was already familiar with the daughter's reaction. In the past it happened now and then that the daughter woke up in the middle of the night, but she then adjusted her sleeping position and went right back to sleep. The daughter had developed anxiety because of the increased number of arguments and was afraid of not being loved and/or being abandoned.
The solution to this problem was painfully obvious: the parents should treat their daughter with love and affection and more importantly talk to her to regain her trust. What's valuable about this story is that the daughter was able to act out her anxiety and did not suppress it. It could also could have turned out very differently: the daughter suppresses her anxiety. In later life she gets to know a man and everything is as it should be. Out of the blue, her husband has to spend more time away from home on business trips.

The old suppressed pattern: ‚none loves me. I have to ensure that my husband does not abandon me...' returns with a vengeance n the form of panic attacks which no one can explain at first until perhaps a therapist recognises and processes the cause with the beleaguered woman.

The quintessence: children are extremely sensitive and can sense vibrations in a room as well as the general family atmosphere. Most of the time, they perceive such things without talking about them and draw childlike conclusions on their own.

To support their daughter I gave her mother the ‚Basic Trust' Chip. Over the past years I have notched up many successes using this chip with my clients. It is also used in a psychosomatic clinic in the meantime. The chip gives the user the feeling of being wanted, welcome and accepted. This, in turn, allows the feeling of Basic Trust to emerge.

Mental balance using the ‚Basic Trust' Chip

I was in a very bad mental state before my attending doctor recommended the ‚Basic Trust' Chip to me. In fact, my condition was so bad that I was in in-patient care at a psychosomatic clinic. My mood was depressed and my thoughts were racing. I had great difficulty slowing down and remaining calm. I was very suspicious of people and created a lot of stress for myself. That's why it was constantly difficult and exhausting for me to cope with everyday life.

After a few weeks of therapy, my doctor gave me the ‚Basic Trust' Chip to support my development. The chip is, in a way, comparable to a CD, and I am, essentially, the CD player. For the application I was meant to look for a quiet place, take the chip in my hand and think of something that reminds me of the feeling of ‚being allowed to let myself go'. Without delay I tried it. When I imagined this feeling, I thought of a fellow patient. He gave me the feeling of calmness and balance. Someone who believed in me and my capabilities.

I have been wearing the ‚Basic Trust' Chip in my bra on the left side above my heart since then. Just the thought of this chip containing my and his positive, strong and serene energy gives me stability and peace. Knowing that this chip and this power are always at my disposal gives me even more power. Power that is in me anyway but is awakened by the chip. I can only recommend this product.

Best regards and sincere thanks, S. G.

The energy transfer

I was visited by nightly visions more and more frequently. I was meant to charge objects with energy. For that reason I manufactured, among other things, specially treated granite plates to be used as desk pads. The idea was to help people who spend a long time working on their computers to maintain their energy and to boost their ability to work for longer periods of time without losing concentration.

Ute, my partner at that time, had one of the granite plates under her desk. One day I got a call from a customer who wanted me to perform a building biological health check. His therapist had detected exposure to electrosmog and I was now supposed to ascertain where exactly it was coming from.

A short while previously I had purchased a special device to measure people's exposure to electrosmog. I wanted to get to know this device and its functions and adjusted the frequency to electrosmog pollution. Ute was sitting at her desk and working on the laptop which was on top of the granite plate. I selected her as my test person and determined that the device revealed no exposure. For that reason I asked her to use my laptop which was not on top of a granite plate. Now the device clearly showed there was exposure to electrosmog. I had Ute go back to her laptop - again no exposure! It was not until we removed the plate under her laptop that exposure to electrosmog was detectable.

My mind was abuzz and my thoughts were racing when I started to put two and two together.

I came to the conclusion that the vibrations of the energetically charged plate had either reduced the electrosmog or influenced the organism so positively that it remained stable making any electrosmog exposure undetectable.

When using my mobile phone a short while later, it dawned on me that mobile phones represent a source of electrosmog. Truth be told, a granite plate is pretty awkward to handle at the best of times so that set me thinking about what format would work for an information carrier so that it could be attached to a phone battery or to a phone cover. A friend of mine in aviation technology produced a printed circuit board onto which I transferred the same frequencies as the granite plate. Initial testing revealed that the printed circuit board worked and had the same effect as the granite plate.

I asked a doctor friend of mine to test the product. He was sceptical and told me in no uncertain terms that the other ‚stickers' on the market he had come into contact with had proven to be relatively ineffective. He visited me at home and carried out initial testing on the printed circuit board. He determined that it worked but remained sceptical. For that reason he wanted to subject the printed circuit board to a longer testing period. He was not yet convinced because his previous experience led him to believe that the plates could not maintain their energy and effectiveness over a longer period of time.

He took the prototype to his practice. After about two months I heard from him again and he told me he was pleased with the product. He asked me to manufacture more of the plates because he was travelling to a conference in two weeks where he would have the opportunity to present the product and have it tested by other doctors and therapists.

I seized this opportunity and produced ten further plates which I then sent to him. I did not hear anything more from him for awhile.

The game-changing mail from the USA

Three months later I received an email from a conference speaker in the USA by the name of Charles Krebs. He provided me with a certificate based on a double blind test with a group of test persons. Numerous products had been tested and my printed circuit board achieved the best results.

In the meantime I had optimised the product and named it the ‚Galileo' Chip.

Dr. Krebs wanted to make my acquaintance when he was next in Germany. Half a year later I had the pleasure of meeting him in Essen. He was very enthusiastic about my improved product and my work as a building biologist so he asked to give a lecture in front of his seminar attendees.

I submitted more chips to Dr. Krebs for further testing. One of these chips is in his mobile phone to this very day. After more than ten years of use, it still works.

Meeting Prof. Fritz Albert Popp

By inventing the chips, I had the pleasure of getting to know many people over time. At this time I was grappling with existence of biophotons and came across the name of Professor Dr Fritz Albert Popp again and again. I wanted to get to know this man in person.

One evening I got a call from a healer by the name of Renzo Celani who wanted to know more about my chips. I was in Greece at the time so had to put him off until my return. I asked him to call me again a week later. I was beyond surprised when he called me one day after my return late in the evening and asked if I would like to meet Professor Popp in Neuss. He was spending time there regularly to carry out tests. Two days later we met up in an old rocket station where Professor Popp had use of the premises. We took one look at each other and got on like a house on fire. He and his son Alexander gave me a tour of the premises and offered to test my ‚Mobile Phone' Chips and energy cards with a new regulation therapy device.

Using this method we were also able to determine that the ‚akury eProtect', the current name of the chip, was able to not only maintain the stability of the body but also to remove blocks and boost the energy level. This finding meant that ‚akury eProtect' could be used for a multitude of issues.

Further examinations using ring cameras and dark-field microscopy ensued and all came to the same conclusion: the ‚akury eProtect' is hugely effective and is able to strengthen the immune system maintaining the stability of the human organism even when exposed to be strong electromagnetic fields.

Rapid improvement with the ‚Basic Trust' Chip

A foster mother came to me with her six-year-old daughter. The girl had not had many opportunities to experience anything remotely like happiness in her first three years in this world: drugs, violence, abuse, hunger... The muscle test immediately indicated that ‚Basic Trust' Chip was required. After the treatment, I checked the device to ascertain if she should continue to use it and for how long. It indicated ‚for ever'. So I gave it to the mother and told her to put it in the child's pillowcase. After two weeks the mother wrote to me telling me that the girl had since gone back to sleeping in her own bed . When she emerges from her room in the morning, she does so with her pillow under her arm.

K. M., therapist from Austria

The breakthrough

Despite this positive result, I was still very much interested in having the effectiveness tested using other testing methods. Dr Charles Krebs referred me to the University of Mainz which was prepared to carry out a study with a device recognised by conventional medicine, the electroencephalography.

Excerpt from Wikipedia: Electroencephalography (EEG) is an electrophysiological monitoring method to record electrical activity of the brain. It is typically non-invasive, with the electrodes placed along the scalp, although invasive electrodes are sometimes used, as in electrocorticography. EEG measures voltage fluctuations resulting from ionic current within the neurons of the brain.[1]Clinically, EEG refers to the recording of the brain's spontaneous electrical activity over a period of time, as recorded from multiple electrodes placed on the scalp.[1]Diagnostic applications generally focus either on event-related potentials or on the spectral content of EEG. The former investigates potential fluctuations time locked to an event, such as ‚stimulus onset' or ‚button press'. The latter analyses the type of neural oscillations (popularly called ‚brain waves') that can be observed in EEG signals in the frequency domain.

When determining the criteria for the study, which was planned for a period of six months, it became clear to me that the selected approach was, indeed, very elaborate but was most certainly the right method to define the efficiency of the chip. We carried out a number of trials on some test persons to finalise the testing procedure. While doing so, we ascertained that after a period of a 90 seconds using a smartphone all test persons displayed an increase in gamma waves on the monitor. This state persisted for over an hour until the brain returned to its normal state (the area of beta and alpha waves.) What exactly did that mean? When the brain is working in the gamma wave area (the area between 40 and 100 Hz), the brain is working at maximum capacity and has to provide an enormous amount of energy throughout this phase. Using the ‚akury eProtect' on the phone, the brain remains in the area of beta waves although the test persons also had to complete complex tasks during the telephone call.

The conclusion is that the test persons with the ‚akury eProtect' on their smartphones could solve complex tasks with less energy input more easily. Their concentration was significantly better and they make fewer mistakes than the test persons without the ‚akury eProtect' on their phones. After more than 30 test persons took part in the standardised testing procedure twice, we were in a position to draw the conclusion previously mentioned conclusion.

For some scientists the results were so astonishing that they were unanimous in their belief that each and every person who uses a smartphone should be provided with the ‚akury eProtect' in their own interest.

Energetic carrier materials, similar to the ones I have developed, made by other manufacturers are readily available on the market. The ‚akury eProtect' is, however, the first and only product for which there is scientifically proven and verifiable evidence of its effectiveness.

The success of the study was the impetus to make further changes to the AkuRy GmbH which I founded in 2007. In 2017 the whole concept was improved in cooperation with a renowned design company. Part of the concept was a newly revised branding, a new website and a newly designed logo. At that point the ‚Galileo' Chip had already been renamed the ‚akury Phone' Chip but the dawn of the age of smartphones made the development of a stronger chip called ‚akury Duett' necessary. The ‚akury Duett' is now known as the ‚akury eProtect'.

The individual areas now combine to make an overall well-rounded concept. Just like in the IT sector, building biology and energy work represent the software whereas the akury Information Chips are the hardware, as it were. The offer is completed and rounded off by the seminars about self-care medicine as it pertains to the expansion of consciousness with the objective of being able to manage your own energy, power and well-being.

The foundation of the akury Institute

This unique and well-rounded overall concept has not found any imitators up to now. Recognizing this was truly ground-breaking for me as this valuable knowledge cannot possibly the lost to the world. It is meant to be shared.

For that reason I founded the akury Institute to train, educate and treat people preventively in a holistic sense i.e. taking account of aspects (physical, mental, emotional, the social surroundings and the way of life) as well as providing assistance.

The guiding principles of the akury Institute are the respect and esteem of each and every person just the way they are, a caring and open way of interaction and the realisation and experience that healing largely depends on a trusting relationship between the individual in need of help and the health consultant.

The objective of this support is the best possible healing of the person in need from a holistic point of view. One truly essential component of this is how we strengthen the individual's personal competence in terms of self-care medicine.

The Institute sees itself as a supporter and companion in the scheme of ‚helping others to help themselves'. People receive the necessary help with the goal of becoming more and more capable of assuming responsibility for their own health. The Institute consists of three areas: the building biology, the seminars and the energy work.

The meaning of „going online"

I found a retreat house in the Allgäu near the town of Kißlegg where I could give my seminars. The participants were all so satisfied with the seminar that they immediately registered for the akury II seminar. It was of great importance to that me each and every participant went home at the end of the seminar feeling that they were taking something valuable with them. Something they could implement immediately in their everyday lives. With the aid of many practical exercises, some of which also take place outdoors the participants increasingly become aware of their own bodies as the most sensitive instrument they can employ to sense and recognise what is good for them or not good for them. Finally, each and every participant is in a position to connect with the Divine energy to manage their own energy, power well-being at the end of the seminar. I have coined the term ‚going online' for this process.

When considering how exactly I could verify if someone is in fact really ‚online', I came up with a very simple and easily understandable method.

Each participant receives a glass of water to be held in their right hand. The energy of the water as well as the energy of the participants are then measured. Afterwards the participants ‚go online' according to the ritual I developed and let the Divine energy flow from there left hand through the body to the water in the glass.

If a participant had an energy, for example, of 10,000 Bovis units before going online and the glass had an energy of 5,000 Bovis units, we can expect the participant as well as the water to exhibit significantly higher amounts of energy subsequent to the energy transfer. As a rule, participants and the water exhibit energy levels between 20,000 Bovis units and 25,000 Bovis units subsequent to ‚going online'. In the event of a glass of water having the same energy as the participant as detected before ‚going online', it can be concluded that participant merely transferred their own energy to the water and was not ‚online'. Some participants were so touched during the ‚online' period that it was difficult for them to interrupt the energy flow. Many recalled perceiving so much love and bliss in the flow of Divine energy that they would have preferred stay there permanently. Long after the ritual, some participants remained in silence and enjoyed the bliss. No one could fathom or believe simple it is to in contact with the Divine energy.

In my opinion everything has to work simply and be understandable for everyone. the more complicated is to explain something, the more sceptical I become. What is simple, is good!

I'm glad I took the step

When I heard about your energy work, it piqued my interest. What on earth could these seminars be about? I asked some participants about their experiences and received unanimous positive feedback. So I decided to try energy work.

Today I am happy I took this step. The seminar content was valuable and enlightening for me. It was for everyday use and depicted in a way best described practical. My own initial progress spurred me on to persevere.

The fact that I met all (!) the participants of the first seminar at the follow-up seminar, akury II, speaks for itself and tells you all you need to know about the quality of the work. The warmth, openness and wonderful hospitality that you experience in the akury Institute are also something very special. I recommended the akury seminars to friends and their positive feedback confirmed that it was the right decision.

K. L., Münster

Testimonial of a participant Mind over matter

As an energy worker and the developer of akury Information Chips I was often invited by Professor F. A. Popp (a German biophysicist who has researched the so-called biophotons since the seventies) to his Institute in Neuss where I took part in research and tests along with other healers. One day a delegation Belgium turned up with freshly extracted tumour cells. First of all, the biophotonic radiation of the tumour cells was determined. Afterwards I along with a Belgian healer were supposed to influence the cancer cells positively using only our minds. This was to take place over a period of 15 minutes. The cancer cells were encased in a lead capsule. The biophotons were measured throughout the procedure. The test showed that we had changed something for the better. This change, however, did not remain stable. The biophotonic radiation of the tumour cells returned to the original state after 30 minutes.

We had still managed to prove that mental powers, be they positive or negative, were not stopped by the lead casing and were able to influence a substance. To have any kind of effect at all, it was important to be able to resonate with tumour cells and to bring about a change using the mind. Maybe someday we will be mentally in a position to heal many illnesses in this way.

Not just the ambience is of great importance at my seminars but also the ‚participants' well-being and satisfaction in terms of culinary delights. My loving wife Doris, ‚my angel' ‚treats the participants to her culinary skills throughout the entire seminar. It is not unusual for a participant or two to enjoy their time with us so much that they would stay if they could. It really is very special, loving and powerful energy that blossoms throughout the seminar. This energy is something some participants have never really experienced before. Nowadays the seminars take place in our home in Annelsbach. The property is 1500 sq metres in size. I have equipped the property with a number of stone circles to boost the energy level making this place a remarkably powerful spot.

Experts take notice

At some point I received an invitation from the renowned Dr. Dietrich Klinghardt to come to Kirchzarten to give a lecture about ‚akury eProtect'.

Dr. Dietrich Klinghardt is a doctor, scientist and teacher. The methods he has developed for diagnostics and therapy are a living system in which new experiences from his medical practice, from science and clinical research are constantly incorporated and thus continuously expand his teaching.

Excerpt from www.klinghardtacademy.com: Dietrich Klinghardt (*1950 in Berlin) studied medicine in Freiburg and has been working as a physician in the USA since 1982. Early on, he specialised in the treatment of chronic diseases. He was not only interested in the appearance of a disease, but also began to research its cause. He quickly came up against the limits of conventional medicine and developed a wide range of alternative methods. Over the years has developed its own forms of diagnosis and therapy based on kinesiology (autonomous regulation test, psycho-kinesiology, mental field techniques), which are now known as the ANK - Applied Kinesiology according to Dr. Klinghardt - are used in medicine.

Dr. Klinghardt used the chip for therapeutic purposes and wore it on his body during his long flights to mitigate the effects of jet lag. When I had successfully completed my lecture for the audience of over 75 participants, a doctor piped up and explained to the participants that she did not think much of products such as the ‚akury eProtect'. She went on to say that there were now a number of equivalent products on the market and she couldn't even begin to tell which ones were suitable or less suitable. Finally she added, she had developed her own method of protection and blessed her food each and every day.

Without even a moment's hesitation I confirmed that rituals of this kind certainly had a very good effect. Full of conviction I added, we would not need doctors, therapists, medicines and food supple-

ments, in fact, we would not even need to be here, were we connected to the Divine energy every second of our lives. I looked into the astonished faces of the audience who were hanging on every word I uttered and could barely wait to hear what I was about to say next. The information just flowed from my mouth and for a moment I even noticed that I was just a mouthpiece for a higher authority. I didn't think about what I was saying. I just relayed the information I was being given:
‚God is in us, but we are not always with God.' That is the core point: ‚If we were always in the Divine energy, we wouldn't need all this. But as we are not connected to the Divine energy every second of our lives, I have developed these products that are meant to help us to make life easier on a day-to-day basis.

After sharing my position, thunderous applause erupted. Some participants even took to their feet. After the lecture I was surrounded by a crowds of people who wanted to buy my products. Within a very short time everything was sold out as I had, quite honestly, not been prepared for such a rush. That evening I also met a doctor who was very interested in selling my products. Her name was Dr. Doris Goldschmidt and today she is my wife.

My lectures for expert audiences have gained me a very good reputation as a building biologist and bioenergeticist over the years. My lectures have been well received and have already led many conventional physicians to review their preconceived opinions about alternative medicine and energy work.

An amazing experiment

While testing, I made the acquaintance of an alternative practitioner who specialises in dark field microscopy. I was curious and wanted to know how it worked. I made myself available for a self-test. For the blood test in the dark field, she took a drop of blood from my fingertip and examined it with a microscope with magnification factor of up to 1000 times. Thanks to state-of-the-art video technology, I was able to follow the examination on screen and got to take a look at the fascinating world of my own blood. This method impressed me. She quickly interpreted my haemogram. It showed many rouleaux. Rouleaux are clumped erythrocytes that can form when stressors are too strong, such as electrosmog or heavy metal exposure. She carried out the examination three times without any major disparity. I then asked her for a ten-minute break, which I wanted to use for my own personal experiment. During this time I meditated and visualised the clumps dissolving and my blood flowing normally. Then I asked her to re-do the test. The slides from the previous examinations

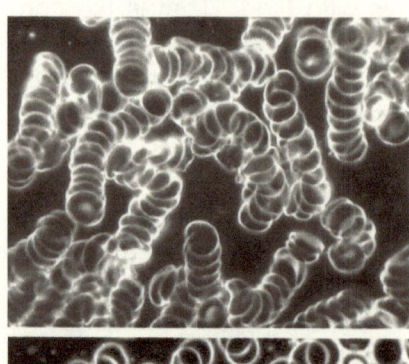
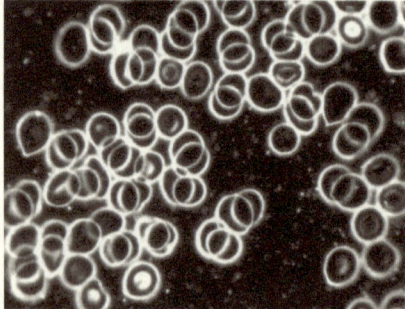

Haemogram before and after meditation

were still on the table. The results of the test were so surprising to her that she asked me if I had taken any medication during the break. My blood was normal and showed no lumps. I explained to her my procedure, which she, in turn, found very plausible. Now I felt the impulse to have her re-examine the slides in front of her from the first round of tests. This finding impressed us both greatly because even these blood samples, which previously showed clear signs of clumping, no longer showed any rouleaux. At first, we were unsure as to how this phenomenon had come about.

Only later did quantum physics give me a plausible explanation for this. Everything is connected and whether a part of me is inside my body or outside my body, as was the case of drops of blood on the sides, does not matter. The moment I influence a particle positively or negatively, it has an influence on the whole system irrespective of time and space. In quantum physics this phenomenon is called ‚entanglement'. Over the course of my seminars I always tell this story to my participants and ask them for the inherent message.

Apart from the realisation that ‚everything is one', there is a lot more to this story. Strictly speaking, it means I can control my own state of mind with the power of my thoughts! I very often use the visualisation techniques with ill people. Correctly applied, it can support a therapy enormously and set the healing process in motion or even accelerate it.

The birth of akury Information Chips

In autumn I drove to Saalfelden in Austria. Dr. Charles Krebs's health was very bad and he could hardly walk up the stairs to his apartment. I had the deep desire to help him and took a blank chip with me. Using a combination of empathy and intuition I perceived which frequencies would be helpful for his recovery and transferred them to the carrier material of the chip. After dinner, I handed the informed chip to Dr. Krebs and invited him to wear it overnight. Dr. Krebs came down the stairs the next day like a new man. He asked me what I had done with him because he was feeling noticeably better. He asked me if he could keep the chip. I said yes and that was the last I heard from him for a while.

After more than half a year I received an email from Dr. Charles Krebs in which he told me with great verve he was achieving incredibly good results balancing and stabilising muscles Not only that but he was also saving a lot of time with his patients with this chip. I improved and extended the frequency spectrum for the chip and sent the new prototypes called ‚Muscle' Chip to the USA.

Today's akury Information Chip

At the next meeting with Dr. Krebs, we spent several days wondering which akury Information Chips would be most important for the human organism.

Dr. Charles Krebs gave me the necessary information to manufacture these chips. Together we developed the first three information chips, which we named ‚TA-harmony' (Thinking Advance), ‚O2-harmony' and ‚AWYN' (All-What-You-Need). A few weeks later the chips ‚LEA' (Life Energy Activator) and ‚FEH' (Five Elements Harmonization) followed.

This marked the beginning of a new era in information medicine. Since 2011 I have developed well over 80 topic-related akury Information Chips, which are now recognised and used by doctors and therapists worldwide. I have chosen ‚Bioenergetic Information Management' as the superordinate term for the akury Information Chips.

Part III

Working with the akury Information Chips

How are the akury Information Chips developed?

The creation of a chip is very complex and requires a systematic approach. The impulses for the development of a new akury Information Chip usually come from the users. First of all, I deal with the topic of the chip, examine the complex connections and make precise notes. As I know that everything that vibrates is also resonant, I find the resonant frequencies in the second step. In the third step the resonant frequencies are merged much like in the creation of a software program. (It is comparable to an orchestra in which all the instruments are in harmony to form a symphony.)

Whenever I create a new akury Information Chip, I always tune into the field of consciousness (morphogenetic field) to obtain more important information required for the information chip.

For example, some time ago a therapist asked me if I could create a chip to stabilise the circulatory system. When I created this chip, I also received the following information which was essential for optimal function: In addition to the circulatory function, rhythm is also important. Our lives are marked by a multitude of rhythms that impact us. They impact us most noticeably when when they are out of sync, for example, breathing rhythm, heart rhythm, biorhythm, sleep-wake rhythm or work rhythm. Last but not least, we are subject to the rhythm of the seasons and the planetary constellations. In the fourth step, all this information is imprinted on the carrier material in the form of frequencies and ultimately programs. The carrier material consists of 15 different substances, which ultimately serve as a vibrating data memory like a USB stick.

The information chip is then ready and subjected to extensive and thorough testing before it goes on sale. A team of approx. 10 therapists test the newly created information chip over a longer period of time in practical use. The above mentioned example is the information of ‚Healthy Rhythm' Chip.

Who the akury information and the, akury eProtect Chip are helpful for

Provided they have a certain mental strength and imagination, everyone could, of course, manage their own strength, energy and well-being. This does, however, require some training if the person is not a natural. In order to make life a little easier for people, I have developed the akury Information Chips according to the quantum physical laws of entanglement and resonance. In a gentle way, the desired frequency patterns are provided to the client's information field via the chips in order to trigger the necessary stabilisation or healing processes in the body. In addition to the activation of the self-healing powers to cope with health problems, there are many other possible applications in our everyday life:

- Maintaining the ability to concentrate

- Achieving optimal performance

- Peace and serenity in stressful situations

- Strengthening mental strength

- Faster regeneration from the stresses of everyday life

- Faster relaxation during rest periods

- Deep and restful sleep.

Nowadays, we are all exposed to a variety of stress factors caused by radio masts, computer work stations, smartphones or our fast-paced lifestyle. This is where the akury Information Chips come in: they can support our immune system and thus keep our bodies stable on all levels. The right akury Information Chips are also helpful in the event of high stress, concentration disorders, lack of energy, burnout, depression and chronic fatigue. In the meantime we boast a selection of over 80 topic-related akury Information Chips. There are thematically arranged sets and many users many users compile their own bioenergetic medicine cabinet according to their own needs.

Information Chips for a multitude of applications.

The application areas of the chips are various.
Many therapists use the akury Information Chips as diagnostic instrument and to support a therapy among other things.

The akury Information Chips are suitable for all people regardless of age, gender or profession. Their application is quite simple: you select the chip for the desired subject area and wear it on your body, for example in your shirt or trouser pocket. The chips can also be easily combined with each other. Unlike medications, where care must be taken to ensure an exact dosage, the application period of the akury Information Chips is unlimited. The body only absorbs the vibrations from the chips it needs to replace disharmonious information with harmonious information. Once disharmonious information has been replaced, the chip does not have anything to resonate with.

What are the advantages of akury Information Chips?

The advantages of the akury Information Chips are clear and can be enumerated in six points:

1. They are easy to use because they are simply worn on the body, i.e. they can be attached to clothing or put in a pocket.

2. They can be used according to need, topic and situation at hand.

3. By their natural principle of action they support the self-healing powers of the organism.

4. The akury Information Chips are not tailored to one person and can therefore be used by the whole family.

5. They do not get used up and and do not consume themselves, even if they are exposed to very strong electromagnetic fields. They retain their effectiveness for at least 10 years.

6. They act in a very gentle way.

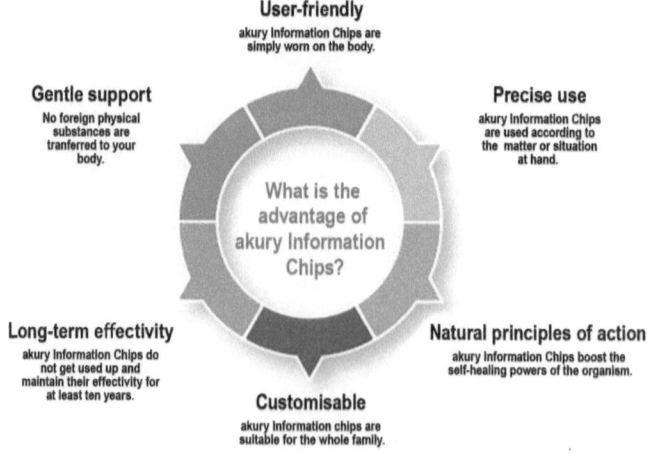

The scientific verdict on the effect of akury Information Chips

The information chips consist of a thin carrier foil in the format of a rounded square of approx. 20 x 20 mm. A mixture of carrier substances is applied to it, which is either informed or energetically charged depending on the effective range.

I handcraft and check each and every akury Information Chips before they are delivered to the customers.

In the case of ‚akury eProtect', the positive effect was proven by an extensive study by the University of Maine. As all akury Information Chips are manufactured according to the same principle and procedure and differ only in the frequency patterns, it follows concluded that all other akury Information Chips are also effective. Individual case studies with the EEG confirmed this logical conclusion.

In addition to the study, there is an expert opinion which was prepared by an independent institute (IIREC) in which the effectiveness of the ‚akury eProtect' was again confirmed.

The akury Information Chips are already being used by a large number of therapists worldwide. The information chips can be used in many different ways and are used by therapists for diagnostics as well as to support the therapy.

A network of therapists is currently under construction. This will benefit those seeking help as they can find a therapist by entering the postcode.

I usually get my inspiration for a new information chisp from the therapists and customers.

Over the years, I have developed a multitude of possibilities and optional combinations to use the information chips in the most meaningfully way possible. This has occurred in close cooperation with the users and has been based on their experience from daily use. Consequently, the akury Information Chips, which are there to help us make our everyday lives easier, can be used by everyone. I have also described these akury Information Chips as ‚chips for everyone' as they contribute to solving a large number of people's problems today. They include, for example, the ‚Deep Restful Sleep' Chip. The majority of people today feel ‚plagued by stress'. The hustle and bustle of everyday life is causing them more and more trouble and can even significantly impair their health. Many process unfinished tasks in their mind when lying in bed. The combination of the anti-stress ‚Relaxation' Chip and the ‚Quick Recovery' Chip create the basis for deep relaxation.

Another almost indispensable combination is the ‚Concentration' Chip used with the ‚Memory' and ‚Keep Cool' Chips. Who doesn't want to be able to commit themselves to their tasks with pin-sharp clarity, calm and serenity? Or to work effectively and maintain an overview in tricky situations? Or to be able to make clear decisions when it really counts?

Amazing results.

Yes, we have achieved amazing results with the following chips

The ‚**Concentration Chip'**: amazing grounding effect and support for college students. Worn 12 hours a day in clothing.

The ‚**Deep Restful Sleep Chip'**: this chip has helped people with intestinal problems who constantly have to get up at night to use the bathroom. This frequency chip calmed their disrupted sleep pattern and they now enjoy an undisturbed night's sleep.

So far these have been the most successful chips. We are still in the process of investigating others. I will provide with more information in due course.

Philip Rafferty, kinesiologist from Deloraine (Australia)

More balance and freedom while enjoying more concentration and energy too.

The use of akury Informaton Chips still does not fail to impress me. They help me in difficult situations. They help me balance my life and boost my performance. For me it was clear as day that I could leave the extremes of either ‚going through life with the brakes on' or ‚being permanently on turbo' behind me with the help of the chip. It was possible for me to make my everyday life more balanced and free while maintaining concentration and an abundance of energy.

Inge Berg, Höchst/Odw.

Calming ‚magical chip'

I would like to thank you from the bottom of my heart for your honesty, help and loving care. In the course of a pretty tough crisis, I got my hands on the ‚Keep Cool' Chip. That was my calming ‚magical chip'. It helped me a lot to calm down.

R.S., Höchst/Odw.

Not short of amazed at how well I got through the surgery.

I wanted to give you feedback on the pain tapes. I attached a pain tape half the way between the navel and the scar immediately after my caesarian section. Even within the first days of the delivery I took only half the offered dose of painkillers. In the following days I reduced it to one tablet in the morning and one in the evening.

I recovered extremely quickly and needed virtually no support from the nursing staff. All in all, the doctors and nurses were quite sceptical about the tape at first. while remaining positive in their professional way. They were amazed at how well I got through the surgery and how I was doing in the days that followed.

Vicky Bobe, kinesiologist from Hilzingen.

Pain and Fibromyalgia

Yet another very popular chip is intended for the relief of pain and is called ‚Relief'. The soothing chip helps to alleviate or dissolve the painful blocks to energy flow caused by physical and/or psychological. This can lead to a greater sense of well-being.

In terms of pain, I have developed a special chip for people suffering from fibromyalgia syndrome. Fibromyalgia is described a range of complaints where the affected party suffers primarily from strong muscle pain throughout the whole body. The term literally means ‚fibre muscle pain'. The chip has already been tested on several fibromyalgia clients and has shown amazing alleviating characteristics.

The relief that went unnoticed at first

I gave the ‚Fibro' Chip to a good friend who suffers from fibromyalgia for testing. There were times when she was in so much pain that she was not even able to get out of bed in the morning, let alone go about her daily routine. After about six weeks, I checked in with her to ask if the chip had already shown any effect. She said that the chip had not brought her any noticeable relief so far. A few minutes later, however, she proudly told my wife that yesterday she had been working in the garden all day for the first time in ages and that she had reduced the strong painkillers by half. I asked her when these changes had come about. She took a moment to ponder my question. After a short while she said slightly embarrassed: ‚actually, since I have been wearing the ‚Fibro' Chip.

Time and time again I have experienced that most people are sceptical about any modern method and want to try out the akury Information Chips before buying them. For a long time I thought about how best to build trust with prospective customers and decided to develop two tapes. One is an alleviating tape and the other is with the vibrations of the ‚Deep Restful Sleep' Chip. Both can be used for the customers personal ‚trial period' and as they are considerably cheaper, they provide the user with a good opportunity to test their effect before committing to the purchase of an akury Information Chip.

There is a further chip combination intended for people who have to withstand the high demands of everyday working life. It does not matter whether the work is physical or mental. These include, for example, hard-working people in the construction industry, professional athletes, managers, students and pupils too.

The most important akury Information Chips to consider here are ‚Quick Recovery', ‚Mental Strength', ‚Reboot Kick/Restart'.

The ‚Quick Recovery' Chip is used for a quick recovery after a mental and/or physically strenuous phase. Furthermore, it is meant to contribute to the best possible uptake and utilisation of oxygen as well as to faster detoxification and removal of waste products from the organs and blood. It is suitable for athletes and people who have to meet high mental and physical demands.
In combination with the ‚Mental Strength' Chip its effect is boosted whether you are a recreational athlete, a professional athlete, student or top manager. Mental strength often gives you the edge you need. This chip promotes your capacity to focus and concentrate in order to be able to achieve the optimal performance. It promotes your capacity to visualise and imagine in order to think and act beyond your own limits.

Another much sought-after akury Information Chip is the ‚Reboot Kick/Restart'. Quite often people come to me feeling depleted and listless barely in a position to cope with their daily work. Usually they are on the verge of a burnout or even in the initial stages of a depressive phase. This stage is comparable to a computer whose batteries are just about empty. In this case you need an impulse to re-start. This is also the case with people whose batteries are almost empty and who want to get out of phases of listlessness or even deep depression. The function of the ‚Reboot Kick' Chip is to provide necessary impulses to liberate the individual from this dilemma.

> **The magic power of the ‚Reboot Kick' Chip.**
>
> I was totally exhausted and no longer able to go about my daily work. Lately, I hadn't even been able to get out of bed in the morning. I didn't have any pain indicative of a serious illness, but I was so worn out and listless that I could barely manage to get on my feet.
>
> One day I came across the ‚Reboot Kick' thanks to health practitioner. It literally catapulted me out of my lack of energy and drive. The very next day I was able to go shopping by myself and two days later I was able to take part in my godchild's first day at school.
>
> I am so grateful that such a thing exists.
>
> C. F., Überau

To my own surprise, the akury Information Chips are even successfully used with animals. I receive the confirmation again and again that the animals react particularly well to the akury Information Chips.

> I have been working as an animal health practitioner/psycho-kinesiologist for animals in my own practice for 13 years. For approx. 4 years I have successfully been using the akury Information Chips as part of the treatment of animals. In particular, ‚Basic Trust', ‚Keep Cool', ‚NHN', ‚SERC', ‚EMO' and ‚FA'. As we do not need to expect anything resembling a placebo effect in the area of veterinary health care, it never fails to surprise me how we can get positive reactions from the animals over and over again. Thank you, Mr Wenner, for your work.
>
> Nicole Nek, animal health practitioner from Wiesbaden

How akury Information Chips are used

The use is quite simple: you select the chip for the desired effect and wear it on your body. That way the bioenergetically effective frequencies are provided o the organism while wearing. The chips can also be easily combined with each other.

My tip for beginners is simply to try it out. The goal should always be to manage one's own energy, strength and well-being. You can't do anything wrong because the body only absorbs the vibration it needs.

I recommend first experimenting with two or three akury Information Chips. If someone is not satisfied, they can return the chips within three weeks.

My Vision: the future of bioenergetic information

More and more people all over the world are realising that, in addition to conventional medicine, there is now a wide range of alternative options for maintaining or recovering health.

Although the term ‚Bioenergetic Information Management' (BI), which I coined, is still very new, it will become an everyday term in the next twenty years. The use of akury Information Chips will be as much a part of our daily lives as brushing our teeth.

Heiko Wenner and a selection of akury Information Chips

Physical medicine will surpass biochemical medicine. People will learn to manage their life energy themselves.

As an ambassador of the ‚Bioenergetic Information Management' I invite you to visit one of my lectures or energy seminars. Further information can be found on the websites

www.akury-energiearbeit.de or www.akury.de.

Conclusion

One evening I was sitting in quiet contemplation on our balcony gazing out over the expansive hilly landscape of the Odenwald into the distance and thinking about the course of my life. I became more and more aware that everything in my life had been predetermined. I saw the decisive phases of my life right before my eyes like jigsaw pieces which now come together to form an overall picture. I realised that nothing in my life so far had been in vain. Everything had its meaning and made me who I am today. And everything I have done is a building block of what I will do.

As a building biology health consultant I recognise the problem areas that affect our health in the long run and how in the case of an ill person they can inhibit or even bring to a standstill the healing processes. Harmonising the identified interference fields simplifies enormously my work as a healer. The akury Information Chips I have developed round off the overall concept. They have a supporting effect and contribute to the stabilisation of the bioenergetic system.

In my seminars I, of course, also deal with all three building blocks because they facilitate a fuller understanding of how to manage your power, your energy and your well-being by yourself'.

Acknowledgements

I would now like to thank all the people who have accompanied and supported me in my life. My gratitude goes to my mother, who gave me life despite the adverse circumstances and bad conditions at that time. My father, through whom I learned toughness, perseverance and stamina. My paternal grandparents, who gave me love, respect and a reverence for nature and animals. My long-time partner Ute, who always stood by me in hard times and also supported me financially. My beloved ‚angel' Doris, who supports me and goes above and beyond to make me feel good. I would like to thank all people who support me in the development of my products and stand by my side so I can share them with the world. Last but not least, I thank my Creator, who has fortunately provided me with an abundance of mental and physical strength, so that I can continue to be a good tool for Him.

www.ingramcontent.com/pod-product-compliance
Lightning Source LLC
Chambersburg PA
CBHW031431210526
45464CB00005B/2155